A SEPARATE CREATION

In the dusk of the evening I took a stroll along a chain of ponds, which in this dry country represented the course of a river, and had the good fortune to see several of the famous Platypus or Ornithorhyncus paradoxus. *They were diving and playing about the surface of the water, but showed so little of their bodies that they might easily have been mistaken for water-rats . . .*

A little time before this I had been lying on a sunny bank, and was reflecting on the strange character of the animals of this country as compared with the rest of the world. An unbeliever in every thing beyond his own reason might exclaim, "Two distinct Creators must have been at work; their object, however, has been the same, and certainly the end in each case is complete."

From Charles Darwin's *Journal of Researches into the Geology and Natural History of the Various Countries Visited by HMS Beagle* (1839).

A SEPARATE CREATION

Discovery of Wild Australia by Explorers and Naturalists

Selected and annotated by

Graham Pizzey

CROOM HELM

London and Dover, New Hampshire

PAGE i
Straw-necked ibis. *Threskiornis spinicollis*, in flight.
PAGE iii
White-tailed kingfisher, *Tanysiptera sylvia*.

Publisher's Note

Original spelling and punctuation
have been retained in quotations
from the journals of explorers,
botanists and naturalists.

This edition published by
Croom Helm Ltd.
Provident House, Burrell Row, Beckenham
Kent BR3 1AT, England
Croom Helm USA
51 Washington Street
Dover, New Hampshire 03820, USA

ISBN 0-7099-4025-4

Printed in Hong Kong through Bookbuilders Ltd

Contents

INTRODUCTION

There is no continent like Australia. There is no vegetation quite like Australia's. There are no mammals, birds, amphibians, reptiles and insects like Australia's old endemic fauna. There are no people quite like Australia's Aborigines. Natural Australia is unique: it is so largely because its plants and wildlife were able to develop in isolation for so long—an accident of geography, or geology.

When the great far-southern landmass, known rather oddly as Gondwanaland, began to break up some 50 million years ago, leaving Antarctica behind, its drifting parts mostly came to rest hard against northern continents and fused with them. India, first to separate, drifted north across the Equator to plough into the underbelly of Asia, buckling up the Himalayas and becoming part of that continent. Africa drifted north to collide with what is now southern Europe, so that the evolving plants and wildlife and, at length, peoples, of northern Africa, Europe and Asia Minor have interchanged freely.

South America joined North America and, as had already occurred in Europe, its early marsupials had to contend with a growing horde of placental mammals invading along a central American isthmus frequently much broader than it is today. Only one marsupial, the American opossum, succeeded in turning the tables by invading north America from the south, and surviving.

Among continents, only Australia and New Guinea, forming one landmass until recent times, remained isolated by sea from all the rest. As this greater Australia had drifted north, between 40 million and 20 million years ago, it was surrounded by oceans. Its climates became mostly mild and humid and much of the continent, to the far inland, was clothed in rainforests, freshwater lakes and great intrusions by the sea.

Then, in the mid-Miocene, some 15 million years ago, after the continent reached its present position off south-eastern Asia, climates began to cool and become drier. This trend culminated in periods of extreme aridity over the last two million years, during the Pleistocene ice-ages.

The slow trend to aridity over that enormous ocean of time, 15 million years, put the final stamp of 'Australian-ness' upon the continent. Vegetation, adapting endless strategies to survive in the new, harsh conditions, took on the grey-green 'drabness' of the Australian bush and the 'harshness' and 'strangeness' of hakea, banksia, grevillea and eucalyptus.

Over the last five million years, according to recent thought, grasslands began to spread, and this led in particular to an extraordinary radiation in the family of kangaroos. Probably descended from possum-like ancestors, they now assumed sizes and forms to suit the new grasslands and open woodlands. They ranged from the tiny, primitive musky rat-kangaroo which clung to remnant rainforest pockets and changed little, through a whole suite of wallabies, to the finest surviving marsupial of all, the lanky, arid-adapted red kangaroo of the inland shrub-steppes and grasslands.

Whole assemblages of other marsupials: koalas, wombats and possums; bandicoots, numbats and marsupial 'mice' and 'moles'; native 'cats' and thylacines arose, as did many forms now extinct.

Over time, those families of birds, amphibians, reptiles, insects and arthropods whose ancestors came up from the south with the continent, also adapted to suit new conditions, their form made more diverse by wide climatic swings. Collectively through that long isolation, these 'old endemic' Australians became perhaps the most different animal assemblage on earth.

The Australian continent was an enormous Ark, with life-support systems of its own. But even before that Ark eventually, some 20 million years ago, ran ashore on the outer edge of Asia, it had already come within reach of a small group of animals able to get here under their own power. Flying insects, birds and bats, wind-drifting spiderlings; desperate, starving rodents and small reptiles rafting on logs, and swimming seals and fish, progressively got here on their own. They formed a new element in the Australian biota.

But apart from these, the continent remained inviolate. Between it and the Asian mainland lay an archipelago populated by very different animals: tiger, monkey, deer and rhinoceros—and eventually, by peoples. But a series of deep channels, and in particular one between what became the islands of Bali and Lombok, meant that no matter how low the Pleistocene ocean-levels fell as the vast polar ice-caps locked up more of the world's moisture, the sea barrier between Australia and Asia remained intact.

To get to Australia you still had to cross wide straits of deep water, and that ultimate barrier denied entry to all but those swimmers and fliers and drifters—until the first ancestors of our Aboriginal peoples reached here, by some form of water-craft, perhaps as long ago as 50 000 years, perhaps longer.

That those supremely adaptable people of the landscape made a considerable mark upon Australia there is no doubt: they greatly increased the incidence of fire; they possibly helped speed the demise of the last giant marsupials like the diprotodons and gigantic kangaroos. By introducing yellow dog dingo some 4000 years ago, they probably helped wipe out the thylacine and the Tasmanian devil on the Australian mainland.

But even over 50 000 years, their influence was as nothing compared to the impact of that most brilliantly constructive and destructive animal, European man, when he finally reached Australian shores three or four centuries ago.

Once discovered, European boarding parties swarmed aboard this ancient Ark. They produced a vibrant new European culture, but laid waste, carelessly, ignorantly and often needlessly, much of the richness that 50 million years of shaping by isolation, climatic change and ice-age had worked.

We, the descendants and beneficiaries of that European invasion, may look with mixed emotions upon the human benefits that have followed, and on the reverse side, the cruel destruction of ancient Aboriginal peoples and cultures, and the devastation of the ground-dwelling Australian marsupials, plant associations and now the very rivers and soils of this extraordinary, beloved continent.

Mankind will never see its like again, never again come to a virgin continent of unbroken, 50-million-year-lineage. What extraordinary privilege the first European explorers to the Australian region had, to come to a continent with living systems so wholly *different*, so wholly *new* to Europeans.

I felt it would be of interest to put together excerpts from the journals of some better-known explorers, observers and naturalists as they described their first reactions to the coasts, inland deserts, rainforests and wildlife never before seen by Europeans.

Perhaps, seeing the impact this new land had upon these first investigators, we may better appreciate the specialness and preciousness of this continent we took from the Aborigines, who in many ways used it more wisely, with more reverence and understanding than we have shown.

Each selected journal entry is accompanied by one or more colour photographs of wildlife those explorers knew: those animals (at least the ones that survive) have not changed. Only the land they inhabit. So in this sense the camera enables us to see precisely what the explorers saw.

However, in trying to match modern photographs with historic text, inequalities must occur. I have tried to be accurate as to the species the explorers probably saw, but in some cases the choices are arguable and in others, precision as to particular *races* has not been achieved. For example, the photograph of the chequered swallowtail butterfly is of an Asian race of this widespread insect.

I must particularly thank Mrs Bronwen Johnston, of Mornington, and Dr Norman Wettenhall and Mr Andrew Isles of Melbourne, for seeking references and lending me journals from their collections. Their sympathy and patience are much appreciated. I would also thank my daughter Sarah Pizzey for research of some quotations difficult to obtain.

I thank Mrs Valda Cole of Frankston; Dr Courtenay Smithers, of the Australian Museum, Sydney; Mr Arturs Neboiss, Ms Joan Dixon and Ms Belinda Gillies of the Victoria Museum, Melbourne; Mr Robert Warneke of the Victorian Ministry for Conservation; the Queensland Forestry Department; the New South Wales Inland Fisheries Research Station, Narrandera, and the Tasmanian National Parks and Wildlife Service, for promptly supplying detailed information.

Finally, my publisher, John Ross, and editor, Margaret Taylor, of Currey O'Neil Ross, have my warm appreciation for their kindness to me and care in the production of the book.

Graham Pizzey
December, 1984

Francis Pelsart

In 1629, Francis Pelsart, a commodore of the Dutch East India Company, was on the ship *Batavia* travelling from Holland by the Cape of Good Hope to the Dutch East Indies.

The ship was carried too far east and, on the calm clear night of 4 July 1629, sailed straight into the coral reef protecting one of the smaller islands (now Pelsart Island) in the Abrolhos group off present Geraldton, in Western Australia.

On several of the larger islands in the group, Pelsart discovered a small wallaby, now known as the tammar, *Macropus eugenii*. (Still found in south-western Australia, on Eyre Peninsula in South Australia and on nearly a dozen coastal islands from the Abrolhos to Kangaroo Island.)

The description Pelsart left is the first known of an Australian kangaroo. The strangeness of the new creature to him is self-evident. It was more than strangeness: there is almost an element of superstition in his reaction to it.

As Europeans would for nearly two centuries, he had difficulty sketching precisely the nature of the animal. It is a clear description in its way, and gets across some essential points, but still has the awkwardness of unfamiliarity that characterises all the early descriptions of kangaroos. How could it be otherwise? No animals like this had ever been written about before. There were no appropriate comparisons. In one detail Pelsart was seriously astray: new-born marsupials do not grow on the nipples, but crawl to the pouch from the urogenital opening and then attach themselves to the nipples. Pelsart's understandable error was long perpetuated, and only laid to rest in modern times.

This excerpt comes from Henrietta Drake-Brockman's *Voyage to Disaster: the Life of Francisco Pelsaert* (Sydney, 1963).

On these islands there are large numbers of Cats . . .

On these islands there are large numbers of Cats, which are creatures of miraculous form, as big as a hare; the Head is similar to that of a Civet cat, the fore-paws are very short, about a finger long. Whereon there are five small Nails, or small fingers, as an ape's fore-paw, and the 2 hind legs are at least half an ell [*c.* 32 centimetres] long, they run on the flat of the joint of the leg, so that they are not quick in running.

The tail is very long, the same as a Meerkat; if they are going to eat they sit on their hind legs and take the food with their fore-paws and eat exactly the same as the Squirrels or apes do.

Their generation or procreation is Very Miraculous, Yea, worthy to note; under the belly the females have a pouch into which one can put a hand, and in that she has her nipples, where have discovered that in there the Young Grow with the nipple in mouth, and have found lying in it some which were only as large as a bean, but found the limbs of the small beast to be entirely in proportion, so that it is certain that they grow there at the nipple of the mammal and draw the food out of it until they are big and can run. Even though when they are very big they still creep into the pouch when chased and the mother runs off with them . . .

Tammar, *Macropus eugenii*

William Dampier

Willingly or unwillingly, William Dampier (1652–1715) first came to Australia in a small vessel whose buccaneering crew had mutinied and abandoned the captain.

Needing to repair the vessel in a secluded place, they sailed the *Cygnet* to north-west Australia from the Philippines and careened her in the bay that now bears her name. (Cygnet Bay is in the Kimberleys, about 100 kilometres north-west of Derby, off the entrance to King Sound.)

Here the *Cygnet* remained from 4 January to 12 March 1688, giving Dampier a remarkable chance to examine New Holland, as the Dutch were already calling it. It was perhaps the best look a European ever had of the Australian mainland to that time.

Dampier left the *Cygnet* in the Nicobar Islands, reached Sumatra by canoe and made his eventful way back to England. There in 1697 he published *New Voyage Round the World* which attracted the Royal Society's interest and led to him being commissioned to return to the South Land and report further.

He reached the west Australian coast in the *Roebuck* in July 1699 and spent two months examining the Abrolhos Islands; Shark Bay (which he named); Dampier Archipelago and then the mainland coast at Roebuck Bay, near today's Broome.

In general the report Dampier gave of Australia was unenthusiastic.

More familiar with shorebirds and seabirds, he not only described but illustrated four species. The illustrations are crude but recognisable and have the distinction of being the first Australian bird illustrations. The birds were the red-necked avocet, *Recurvirostra novaehollandiae*; the pied oystercatcher, *Haematopus ostralegus*; the bridled tern, *Sterna anaetheta* and the common noddy, *Anous stolidus*.

... all singing with great Variety of fine shrill Notes

The Blossoms of the different Sort of Trees were of several Colours, as red, white, yellow, &c. but mostly blue: And these generally smelt very sweet and fragrant, as did some also of the rest. There were also beside some Plants, Herbs, and tall Flowers, some very small Flowers, growing on the Ground, that were sweet and beautiful, and for the most part unlike any I had seen elsewhere.

There were but few Land-Fowls; we saw none but Eagles, of the larger Sorts of Birds; but 5 or 6 Sorts of Small Birds. The biggest Sort of these were not bigger than Larks; some no bigger than Wrens, all singing with great Variety of fine shrill Notes; and we saw some of their Nests with young Ones in them. The Water-Fowls are Ducks, (which had young Ones now, this being the Beginning of the Spring in these Parts;) Curlews, Galdens, Crab-catchers, Cormorants, Gulls, Pelicans; and some Water-Fowl, such as I have not seen any where besides. I have given the Pictures of 4 several Birds on this Coast . . .

Pied oystercatchers, *Haematopus ostralegus*

OPPOSITE: Red-necked avocet, *Recurvirostra novaehollandiae*

William Dampier

There is probably no better, or worse, example of European unfamiliarity clothing a harmless and not particularly puzzling Australian creature in superstitious humbug than in this description by Dampier of his meeting with a shingle-back lizard, *Trachydosaurus rugosus*.

On being confronted, the shingle-back—as they will—put on its best bluff and no doubt displayed a large area of blue tongue and shocking pink mouth-lining.

Allowing for Dampier's quaintness of expression there is a sense of almost medieval alienation here, quite apart from a very understandable reluctance to eat a smelly lizard. Dampier's reaction was a measure of the adjustment he and all Europeans needed to make to things Australian.

. . . this Creature seem'd . . . to have a Head at each End

[We also saw] a Sort of Guano's [that] had a Stump of a Tail, which appear'd like another Head; but not really such, being without Mouth or Eyes:

Yet this Creature seem'd by this Means to have a Head at each End; and, which may be reckon'd a fourth Difference, the Legs also seem'd all 4 of them to be Fore-legs, being all alike in Shape and Length, and seeming by the Joints and Bending to be made as if they were to go indifferently either Head or Tail foremost.

They were speckled black and yellow like Toads, and had Scales or Knobs on their Backs like those of Crocodiles, plated on to the Skin, or stuck into it, as part of the skin.

They are very slow in Motion; and when a Man comes nigh them they will stand still and hiss, not endeavouring to get away.

Their Livers are also spotted black and yellow: And the Body when opened hath a very unsavory Smell. I did never see such ugly Creatures any where but here. The Guano's I have observ'd to be very good Meat: And I have often eaten of them with Pleasure; but tho' I have eaten of Snakes, Crocodiles and Allegators, and many Creatures that look frightfully enough, and there are but few I should have been afraid to eat of, if prest by Hunger, yet I think my Stomach would scarce have serv'd to venture upon these *N. Holland* Guano's, both the Looks and the Smell of them being so offensive . . .

Shingle-back lizard, *Trachydosaurus rugosus*

James Cook

When Captain James Cook sailed the *Endeavour* into Botany Bay on 29 April 1770, the continent was, despite earlier European investigations, all but unknown.

Not only were its coasts, rivers, mountains and its limitless hinterland still to be charted, but the nature of the land itself: its origins, its kind of trees, flowering plants, birds, mammals, fish and reptiles, were as far beyond the knowledge of the English party as was the far side of the moon.

Were there native peoples on this eastern coastline? Were they giants or pygmies, of Oriental or negroid extraction? How did they live, what were their dispositions? How numerous were they?

Were there tigers or lions in the woods, or only animals like the 'species of cats' (actually the little wallaby, the tammar) the Dutch Commodore Pelsart had reported from the Abrolhos Islands in 1629?

So when on 1 May 1770, Cook made his first 'Excursion in the Country', he saw the coastal woodlands with a wholly new eye.

There is no certainty what Dr Solander's 'small Animal something like a Rabbit' may have been. I have elected to show a southern brown bandicoot, *Isoodon obesulus*. It may equally have been a long-nosed potoroo, *Potorous tridactylus* or a bettong, *Bettongia* sp. or *Aepyprymnus* sp.

Southern brown bandicoot, *Isoodon obesulus*

. . . the Country . . . diversified with Woods, Lawns, and Marshes

This morning . . . we made an Excursion into the Country, which we found diversified with Woods, Lawns, and Marshes. The woods are free from underwood of every kind, and the trees are at such a distance from one another that the whole Country, or at least great part of it, might be Cultivated without being obliged to cut down a single tree.

We found the Soil everywhere, except in the Marshes, to be a light white sand, and produceth a quantity of good Grass, which grows in little Tufts about as big as one can hold in one's hand, and pretty close to one another; in this manner the Surface of the Ground is Coated.

In the woods between the Trees Dr. Solander had a bare sight of a Small Animal something like a Rabbit, and we found the Dung of an Animal which must feed upon Grass, and which, we judge, could not be less than a Deer; we also saw the Track of a Dog, or some such like Animal . . .

James Cook

After a week at Botany Bay, Cook, more familiar with his Australian surroundings, made of the woodlands what later came to be an expected observation by newly-arrived Europeans about the Australian bush: that there was little variety; it was monotonous.

In somewhat striking contrast, Banks and Dr Solander, the 'scientific gentlemen', found such a quantity of (different) plants that Cook later decided to call the place Botany Bay: earlier he had considered the rather less enthusiastic epithet 'Stingray Bay'.

Can it be that the scientists had already begun to appreciate the magical complexity of the Australian bush and coastal heaths, while the enormously competent but matter-of-fact Cook, his mind ever fixed on the main purposes of the expedition, saw it in essentially practical terms?

That contrast, that difference in perceptions, is reflected perfectly today in the opposing, mutually inexplicable views of the land held by developers and conservers.

Be that as it may, Cook was not blind to the beauty of the parrots, among which would certainly have been our gorgeous rainbow lorikeet, *Trichoglossus haematodus*, and possibly the turquoise parrot, *Neophema pulchella*.

Turquoise parrot, *Neophema pulchella*

The great quantity of plants . . . occasioned my giving it the Name of Botany Bay

The great quantity of plants Mr. Banks and Dr. Solander found in this place occasioned my giving it the Name of Botany Bay . . . It is capacious, safe, and Commodious; it may be known by the land on the Sea Coast . . . Rather higher than it is inland, with steep rocky Clifts next the Sea, and looks like a long Island lying close under the Shore . . .

Although wood is here in great plenty, yet there is very little Variety; the biggest trees are as large or larger than our Oaks in England, and grows a good deal like them, and Yields a reddish Gum; the wood itself is heavy, hard, and black like Lignum Vitae.

Another sort that grows tall and Strait something like Pines—the wood of this is hard and Ponderous, and something of the Nature of America live Oak. These 2 are all the Timber trees I met with; there are a few sorts of Shrubs and several Palm Trees and Mangroves about the Head of the Harbour.

The Country is woody, low, and flat as far in as we could see, and I believe that the Soil is in general sandy. In the Wood are a variety of very beautiful birds, such as Cocatoos, Lorryquets, Parrots, etc., and crows Exactly like those we have in England . . .

Rainbow lorikeet, *Trichoglossus haematodus*

Joseph Banks

Sir Joseph Banks (1743–1820), KCB, PC, FRS, was a fortunate man who used his fortune to immense effect.

Among other things his influence brought about the convict colonisation of Australia and supported many subsequent exploring and scientific investigations, notably those of Matthew Flinders and Allan Cunningham. Consulted by the British government on many matters Australian, he influenced the selection of governors and countless other decisions through the colony's first two decades.

Son of a wealthy Lincolnshire landowner, Banks was educated at Harrow, Eton and Oxford, where he studied botany under a lecturer brought from Cambridge for the purpose.

He gained an honorary MA at Oxford and subsequently did such effective botanical work in Labrador and elsewhere that at 23 he was elected a Fellow of the Royal Society. He was to be the Society's president for more than forty years.

A connection with the First Lord of the Admiralty enabled the 27-year-old Banks, at his own expense, to recruit and accompany a scientific party on Cook's famed voyage to the Pacific (1768–71) to observe the transit of Venus from Tahiti, and to further explore the Pacific.

He is said to have spent ten thousand pounds, an immense sum then, to equip and employ staff to record their discoveries: the scientist Dr Daniel Solander, the artists John Reynolds, Alexander Buchan and Sydney Parkinson, and Herman Spöring, Banks's secretary.

Cook generously offered Banks's party the use of his great cabin on the *Endeavour*, though it was one of the few places in the ship where his tall frame could stand upright.

It was Banks, at least as much as Cook, who carried forward the long, irregular process by which the European world gradually awakened

Crimson rosella, *Platycercus elegans*

to the existence of such a land of difference far below the horizon.

The excerpts here are from the facsimile edition of *The Endeavour Journal of Joseph Banks 1768–1771*, vol. 2, edited by the historian J. C. Beaglehole (Sydney, 1962).

Banks's famous account of the shyness of Australian birds was echoed by most explorers. Except on remote offshore islands, Australian birds and mammals were 'shy as deer', as Flinders put it. But few shared Banks's conclusions that the Aborigines were not responsible for this state of affairs, or were 'not very clever in deceiving' birds.

The Aborigines were in fact amazingly skilled bird-hunters, using techniques suited to species and situation: by stalking, by climbing, trapping, snaring, netting, throwing well-aimed stones, sticks or boomerangs; by spearing, or by seizing silently and smoothly from under water. So there was good reason for New Holland crows not to be fools: they either got wary, or got eaten.

Red-tailed black cockatoos, *Calyptorhynchus magnificus*

. . . this extrordinary shyness in the Birds

The Land Birds were crows, very like if not quite the same as our English ones. Parrots and Paraquets most Beautifull, White and black Cocatoes, Pidgeons, beautifull Doves, Bustards, and many others which did not at all resemble those of Europe.

Most of these were extremely shy so that it was with dificulty that we shot any of them; a Crow in England tho in general sufficiently wary is I must say a fool to a New Holland crow and the same may be said of almost if not all the Birds in the countrey . . .

What can be the reason of this extrordinary shyness in the Birds is dificult to say, unless perhaps the Indians are very clever in deceiving them which we have very little reason to suppose, as we never say any instrument with them but their Lances with which a Bird could be killd or taken, and these must be very improper tools for the Purpose . . .

Joseph Banks

In botanical collecting, plant specimens were pressed between sheets of paper in a frame to dry and preserve them.

On the *Endeavour*, Joseph Banks had a continual battle to keep his plant presses from damp. At Botany Bay, on 3 May 1770, he brought them ashore to lay the sheets out in the sun and air them—hence his opening comments.

Banks's encounter with numerous quail in the hinterland of Botany Bay reveals an abundance occasionally mentioned by early settlers. His description suggests that, despite the lateness of the season, his birds may have been stubble quail, *Coturnix pectoralis*, whose size and markings somewhat resemble those of the European bird.

During the 19th century, stubble quail and also brown quail, *C. australis*, and painted buttonquail, *Turnix varia*, were occasionally abundant in undeveloped areas of outer Sydney.

In his great *Nests and Eggs of Australian Birds* (Sydney, 1901–04) A. J. North, ornithologist at the Australian Museum, included the following note from a correspondent:

In the year 1851, at a place known as the 'Red Hill', near Bunnerong Road, ran a small swamp down to Botany Bay . . . where a sportsman could bag, with the aid of his dogs, from thirty to forty brace of stubble quail . . . on any day he chose to shoulder his gun, besides other kinds of game.

They remained in this locality all the summer, and on the eastern side of the old Botany Road, leading to the Sir Joseph Banks Hotel; the remnants of the birds were to be found the following season scattered about on the low lying country . . .

The fruiting tree described by Banks was perhaps the cherry ballart, *Exocarpos cupressiformis*. However, it usually bears its curious small fruit in summer.

I . . . found a large quantity of Quails . . .

When the damp of the Even made it necessary to send my Plants and books on board I made a small excursion in order to shoot any thing I could meet with and found a large quantity of Quails, much resembling our English ones, of which I might have killed as many almost as I pleased had I given my time up to it, but my business was to kill variety and not too many individuals of any one species.

The Cap^tn and Dr Solander employd the day in going in the pinnace into various parts of the harbour.

They saw fires at several places and people who all ran away at their approach with the greatest precipitation, leaving behind the shell fish which they were cooking; of this our gentlemen took the advantage, eating what they found and leaving beads ribbands &c in return.

They found also several trees which bore fruit of the Jambosa kind, much in colour and shape resembling cherries; of these they eat plentifully and brought home also abundance, which we eat with much pleasure tho they had little to recommend them but a light acid.

Stubble quail, *Coturnix pectoralis*

Two-brand crow, *Euploea sylvester*

Joseph Banks

On 29 May 1770, the *Endeavour* anchored in Thirsty Sound, which opens off Broad Sound, midway between present Rockhampton and Mackay.

On the edge of the tropics, this is still a good region for butterflies, and Banks was fascinated when he went ashore. The reference Banks gave for the first species he saw suggests they were blue tigers, *Danaus hamata*. But such a gathering as Banks described has been seen by few living entomologists.

The butterfly whose silvery pupae so impressed Banks was probably the two-brand or double-branded crow, *Euploea sylvester*, which ranges from Torres Strait to Rockhampton.

. . . a butterfly of a velvet black changeable into blue

Insects in general were plentifull, Butterflies especialy of one sort . . . the air was for the space of 3 or 4 acres crowded with them to a wonderfull degree: the eye could not be turnd in any direction without seeing milions and yet every branch and twig was almost coverd with those that sat still: of these we took as many as we chose, knocking them down with our caps or any thing that came to hand.

On the leaves of the gum tree we found a Pupa or Chrysalis which shone almost all over as bright as if it had been silverd over with the most burnishd silver and perfectly resembled silver; it was brought on board and the next day came out into a butterfly of a velvet black changeable into blue, his wings both upper and under markd near the edges with many light brimstone colourd spots, those of his under wings being indented deeply at each end . . .

James Cook
Joseph Banks

After near-disaster on a coral reef inside the main Barrier, Cook's expedition spent part of June and all July 1770 careened in the Endeavour River, near present Cooktown, where the ship was repaired. There, at last, they met the kangaroo.

They were not the very first Europeans to see a member of its family: the Spanish in New Guinea, the Dutch in south-western Australia, Dampier at Shark Bay and others, had that honour. But none had left sufficiently detailed descriptions or illustrations of a large kangaroo to quite prepare the *Endeavour*'s people for so unexpectedly *new* an animal.

In those days before photography and before European artists learned to depict Australian animals as anything less than freaks, it must have been very difficult to come to terms with something so *different* without seeing for yourself.

The Australian zoologists John Calaby and the late Dr Harry Frith have put forward the view, in their book, *Kangaroos*, (Melbourne, 1969) that during their time ashore, the *Endeavour*'s company saw and killed two kinds of kangaroos: the eastern grey, *Macropus giganteus*, and the wallaroo, *M. robustus*.

Hesitation in assimilating the very *idea* of such animals, so out of all previous experience, and the difficulty of finding comparative terms to describe them, comes through nicely in these excerpts from the journals of Cook and Banks, respectively, before they or the European world first heard the local Aboriginal word, 'kangaroo'.

I saw myself . . . one of the Animals before spoke off . . .

Sunday 24th. [June 1770] . . . I saw myself this morning, a little way from the Ship, one of the Animals before spoke off; it was of a light mouse Colour and the full size of a Grey Hound, and shaped in every respect like one, with a long tail, which it carried like a Grey hound; in short, I should have taken it for a wild dog but for its walking or running, in which it jump'd like a Hare or Deer.

Another of them was seen to-day by some of our people, who saw the first; they described them as having very small Legs, and the print of the Feet like that of a Goat; but this I could not see myself because the ground the one I saw was upon was too hard, and the length of the Grass hindered my seeing its legs . . .

To compare it to any European animal would be impossible . . .

[June] 25 In gathering plants today I myself had the good fortune to see the beast so much talkd of, tho but imperfectly; he was not only like a fowl, and after the strictest search no gizzard could be found; the legs, which were of a vast length, were covered with thick, strong scales, plainly indicating the animal to be formed for living amidst deserts; and the foot differed from an ostrich's by forming a triangle, instead of being cloven . . .

The wings are so small as hardly to deserve the name, and are unfurnished with those beautiful ornaments which adorn the wings of the ostrich: all the feathers are extremely coarse, but the construction of them deserves notice—they grow in pairs from a single shaft . . .

It may be presumed, that these birds are not very scarce, as several have been seen, some of them immensely large, but they are so wild, as to make shooting them a matter of great difficulty. Though incapable of flying they run with such swiftness, that our fleetest greyhounds are left far behind in every attempt to catch them. The flesh was eaten, and tasted like beef . . .

Emus, *Dromaius novaehollandiae*

Matthew Flinders

Bold, painstaking and blessed with a rare human warmth, Matthew Flinders (1774–1814) did more than any other to draw the veil of ages from Australia's coasts.

From his arrival in Australia in 1795, to 1800, and from 1801 to 1803, Flinders was mostly at sea, exploring and charting. His vessels ranged from Bass's miniscule *Tom Thumb* to the ageing sloop HMS *Investigator*, which he sailed out from England and in which ultimately rotten, unseaworthy vessel he laid down accurate charts of the continent's vast southern, eastern and northern coasts. That accomplishment remains one of man's greatest sustained individual ventures in discovery.

The following excerpts, from Flinders's great saga, *A Voyage to Terra Australis* (London, 1814), reflect some of the man's alert intelligence and his frequent droll amusement at an animal or a human situation. From the sea-eagles that mistook him for a kangaroo in St Vincent's Gulf to the bush flies that pestered them off Arnhem Land, Flinders recorded it all in lucid, economic detail.

His first excerpt concerns a visit he made in the schooner *Francis* to Preservation Island, off Cape Barren Island in the Furneaux group, Bass Strait, as a 23 year-old lieutenant in February 1798.

Here the ship *Sydney Cove* had been wrecked and the survivors, despatching an ill-fated rescue party to Port Jackson, had made vital use of Bass Strait muttonbirds, *Puffinus tenuirostris*, and Cape Barren geese, *Cereopsis novaehollandiae*, both plentiful on the island. Unused to humanity, the geese could at first be killed with sticks but soon learned fear, as Flinders reports.

The later fortunes of the goose in Bass Strait are worth noting. As sealers, then settlers with stock, arrived on many of the Furneaux group islands, the geese declined.

Cape Barren geese, *Cereopsis novaehollandiae*

They and their eggs were good to eat and the geese competed with sheep for pasture.

Visiting islands in the Furneaux group in 1899 after a century of slaughter, Dr Montgomery, Anglican Archbishop of Tasmania (and grandfather of Field Marshall Viscount Montgomery of Alamein), gloomily doubted 'if there are more than 120 nests in the year'. (Did he know this goose nests in winter, I wonder?)

But perhaps to everyone's surprise, the effects of gradual human depopulation of the outer islands and in recent years, improving surveillance by Tasmanian fauna authorities, helped bring the goose back from the abyss.

Breeding also on southern offshore islands of Victoria, South Australia and Western Australia, from which it migrates to spend summer in nearby mainland areas, the Cape Barren goose again numbers thousands and seems reasonably secure.

The 'bernacle' to which Flinders compares the Cape Barren goose was the barnacle goose, *Branta leucopsis*, which Flinders would have known in his native Lincolnshire. The 'red-bills' were either pied or sooty oystercatchers, *Haematopus* spp.

Cape Barren geese, *Cereopsis novaehollandiae*

. . . it formed our best repasts

Of the birds which frequent Furneaux's Islands, the most valuable are the goose and black swan; but this last is rarely seen here, even in the fresh-water pools, and except to breed, seems never to go on shore.

The goose approaches nearest to the description of the species called *bernacle*; it feeds upon grass, and seldom takes to the water. I found this bird in considerable numbers on the smaller isles, but principally upon Preservation Island; its usual weight was from seven to ten pounds, and it formed our best repasts, but had become shy.

Gannets, shags, gulls and red-bills were occasionally seen; as also crows, hawks, paroquets, and a few smaller birds . . .

19

Matthew Flinders

In February 1798 Flinders also visited Clarke and Cape Barren islands, where he met common wombats, *Vombatus ursinus*. Today the wombat of Bass Strait survives only on Flinders Island. Sealers, shipwrecked mariners and earlier settlers helped exterminate it on Clarke and Cape Barren islands, and on Deal and King islands.

Flinders reported that, unlike their mainland cousins, these island wombats apparently wandered about feeding at any time of the day, easy prey for hungry humans. Had there been an Aboriginal population to keep them wary, they may have survived longer.

Flinders, himself a harbinger of change, was seeing the last of the old natural order in Bass Strait. For 14 000 years, since rising ocean levels at the end of the last Pleistocene ice-age separated Tasmania from the mainland and made islands of the granite-buttressed mountain-chain that formerly united the two, the sea had protected the wildlife of those islands. That age-old sanctuary was abruptly shattered when European ships arrived.

It burrows like the badger...

Clarke's Island afforded the first specimen of the new animal, called *womat*; but I found it more numerous upon that of Cape Barren: Preservation and the Passage Isles do not possess it.

This little bear-like quadruped is known in New South Wales, and called by the natives *womat*, *wombat*, or *womback*, according to the different dialects, or perhaps to the different rendering of the wood rangers who brought the information.

It burrows like the badger, and on the Continent does not quit its retreat till dark; but it feeds at all times on the uninhabited islands, and was commonly seen foraging amongst the sea refuse on the shore, though the coarse grass seemed to be its usual nourishment.

It is easily caught when at a distance from its burrow; its flesh resembles lean mutton in taste, and to us was acceptable food . . .

Common wombat, *Vombatus ursinis*

Matthew Flinders

The Bass Strait muttonbird or short-tailed shearwater, *Puffinus tenuirostris*, still breeds in millions on Bass Strait islands, its world stronghold. It is possibly Australia's most abundant bird.

Shearwaters are a world-wide group of small-to-largish petrels whose fast, weaving flight, often slicing the waves, uses the energy of the wind off the sea to economically cover vast reaches of ocean in ceaseless food-search and annual migration.

Decades of patient investigation by biologists like Australia's Dr D. L. Serventy have made familiar the Bass Strait muttonbird's astonishing annual programme. Leaving the breeding islands in the Strait about the end of May, the birds move east into the Pacific then north past Japan to the Arctic Pacific, where they moult.

Then they move down the American west coast and back across the Pacific to make landfall on the breeding islands in the last week in September. After 're-courtship' and cleaning out of burrows, they go to sea again until eggs are laid toward the end of November. (Flinders was for once in error: they lay only one one large, white egg.)

So in September and early October and from about mid-November on, the hungry apprehensive people of the wrecked ship *Sydney Cove* suddenly had a superabundant source of fresh if somewhat oily meat and, for a short time, fresh eggs.

The Bass Strait muttonbirds seem little diminished today. On many Bass Strait islands the darkening summer evening air is still interwoven by multitudes of returning muttonbirds, back from feeding sorties that often span hundreds of kilometres. Standing to watch the swift, dark, weaving, wheeling forms and hearing their weird brayings, you can hie yourself back in mind to that distant day when Flinders saw it all exactly like this.

Short-tailed shearwater, *Puffinus tenuirostris*

. . . the air . . . darkened with their numbers

The sooty petrel, better known at sea under the name of *sheerwater*, frequents the tufted, grassy parts of all the islands in astonishing numbers. It is known that these birds make burrows in the ground, like rabbits; that they lay one or two enormous eggs in these holes, and bring up their young there.

In the evening, they come in from the sea, having their stomachs filled with a gelatinous substance gathered from the waves; and this they eject into the throats of their offspring, or retain for their own nourishment, according to circumstances. A little after sunset, the air at Preservation Island used to be darkened with their numbers; and it was generally an hour before their squabbling ceased, and every one had found its own retreat.

The people of the Sydney Cove had a strong example of perseverance in these birds. The tents were pitched close to a piece of ground full of their burrows, many of which were necessarily filled up from walking constantly over them; yet, notwithstanding this interruption, and the thousands of birds destroyed, for they constituted a great part of their food during more than six months, the returning flights continued to be as numerous as before; and there was scarcely a burrow less, except in the spaces actually covered by the tents . . .

George Bass

George Bass's whaleboat voyage from Port Jackson to what he called Western Port 'on account of its position relative to every known harbour' in the summer of 1797–98 is still celebrated as a feat of sustained daring.

Matthew Flinders, whose opinion counted in such matters, (he was not on the whaleboat voyage) commented: 'A voyage *expressly* undertaken for discovery in an open boat . . . in which six hundred miles of coast, mostly in a boisterous climate, was explored, has not, perhaps its equal in the annals of maritime history'.

Bass came to New Holland driven by one consuming purpose: to win reputation and wealth by discovery or by trade. That fierce ambition was soon to carry him to an unknown death on a voyage to South America.

Vigorous and alert observer that Bass was, it is sad that at Westernport, his eye to his main purpose resulted in so brief and matter-of-fact an account of the wildlife, which we know to have been rich. The region was also experiencing a particularly dry summer.

But what Bass did write was pithy and accurate. His 'small but excellent' ducks were almost certainly chestnut teal, *Anas castanea*, which still find refuge on the Westernport tidal flats in good numbers, although seldom now in 'thousands'.

Black swans, *Cygnus atratus*, may still be seen there in hundreds, though, as later quotations in this book recall, they were in for rough treatment.

Chestnut teal, *Anas castanea*

The land round Western Port is low but hilly . . .

The land round Western Port is low but hilly, the hills rising as they recede, which gives it a pleasing appearance. Upon the borders of the harbour it is in general low and level.

In the different places I landed I found the soil almost uniformly the same all round—a light brown mould free from sand, and the lowest lying grounds a kind of peaty earth. There are many hundred acres of such sort of ground.

The grass and ferns grow luxuriantly, and yet the country is but thinly and lightly timbered. The gum-tree, she and swamp oaks, are the most common trees.

Little patches of brush are to be met with everywhere, but there are upon the east side several thick brushes of some miles in extent, whose soil is a rich vegetable mould. In front of these brushes are salt marshes.

The island is but barren. Starved shrubs grow upon the higher land, and the lower is nothing better than sandy brushes, at this time dried up.

We had great difficulty in finding good water, and even that which was brackish was very scarce. There is, however, every appearance of an unusual drought in the country.

The head of the winding creek on the east side, which I have marked with Fresh Water in the sketch, was the only place we could procure it at free from a brackish taste. At half-tide there is water enough over the shoals for the largest boat, and within the creek there is at all times a sufficient depth.

There seem to be but few natives about this place. We saw only four, and that the day after we came in, but they were so shy we could not get near them. There are paths and other marks of them in several places, but none very recent. The want of water has perhaps driven them further back upon the higher lands.

We saw a few of the brush kangaroo, the wallabah, but no other kind. Swans may be seen here, hundreds in a flight, and ducks, a small but excellent kind, fly in thousands. There is an abundance of most kinds of wild fowl . . .

George Bass

Following George Bass's incredible whaleboat voyage to Westernport, in October 1798 Governor Hunter despatched the *Norfolk*, under Flinders (accompanied by Bass) to attempt the circumnavigation of Tasmania and finally prove the existence of a strait between it and the mainland.

This they did and Hunter, at the generous recommendation of Flinders, named it Bass Strait.

On 9 December 1798, off north-west Tasmania, the *Norfolk* sighted a small island 'high and steep'. Flinders noted in his journal that it 'appeared to be almost white with birds; and so much excited our curiosity and hope of procuring a supply of food, that Mr Bass went on shore in the boat, whilst I stood off and on, waiting for his return'.

This rock, which they called Albatross Island, is one of only three breeding sites of the white-capped albatross, *Diomedea cauta*, in Australian waters: the others are the Mewstone and Pedra Branca, both off southern Tasmania. No other species of albatross breeds in Australia.

Depredation by sealers early last century must have had an effect on this colony, but the great birds are still numerous and in recent years up to 2000 pairs have bred on Albatross Rock each summer.

No such happy fate awaited the Australian fur-seals on Albatross Island. The sealing rush that soon followed the exploration of Bass Strait wiped them out. Today, after nearly a century of official legal protection, although fur seals still go ashore there, no breeding colony has become re-established on Albatross Island.

Mr. Bass returned . . . with a boat load of seals and albatrosses . . .

Mr. Bass returned at half past two, with a boat load of seals and albatrosses. He had been obliged to fight his way up the cliffs of the island with the seals, and when arrived at the top, to make a road with his clubs amongst the albatrosses.

These birds were sitting upon their nests, and almost covered the surface of the ground, nor did they any otherwise derange themselves for the new visitors, than to peck at their legs as they passed by.

This species of albatross is white on the neck and breast, partly brown on the back and wings, and its size is less than many others met with at sea, particularly in the high southern latitudes.

The seals were of the usual size, and bore a reddish fur, much inferior in quality to that of the seals at Furneaux's Islands. *Albatross Island*, for so it was named, is near two miles in length, and sufficiently high to be seen five or six leagues from a ship's deck: its shores are mostly steep cliffs . . .

White-capped albatross, *Diomedea cauta* (photographed on a New Zealand island)

Matthew Flinders

Returning to England in late 1800, Flinders put a plan to Sir Joseph Banks 'for completing the investigation of the coasts of Terra Australis'. The proposal was immediately successful and in January 1801 Flinders was commissioned by the Admiralty to proceed to Australia in command of HMS *Investigator*.

Reaching Cape Leeuwin in December 1801, he immediately began his great survey. Moving eastward, in February 1802 Flinders entered (and later named) Spencer Gulf in present South Australia.

On Sunday 21 February he went ashore with the ship's master on one of the several islands at the entrance of the Gulf. There they applied a singular and uncharacteristically cruel method of securing a venomous snake for study.

There also, not for the first or last time, they saw the apparently complete lack of fear to be found in island wildlife unacquainted with human beings.

The birds involved in this incident were white-breasted sea-eagles, *Haliaeetus leucogaster*, which still breed on islands of the Gulf.

Despite the blood-curdling description given by Flinders, white-breasted sea-eagles are mostly hunters and scavengers of fish. They do take small mammals, like pademelons and flying-foxes, but hardly one as large as a man.

One suspects this pair on Thistle Island may have been as much curious as aggressive—admittedly a fine distinction to an explorer confronted by an onrushing eagle.

. . . it seemed evident that they took us for kanguroos

Upon the whole, I satisfied myself of the insularity of this land; and gave to it, shortly after, the name of THISTLE'S ISLAND, from the master who accompanied me.

In our way up the hills, to take a commanding station for the survey, a speckled, yellow snake lay asleep before us. By pressing the butt end of a musket upon his neck, I kept him down whilst Mr. Thistle, with a sail needle and twine, sewed up his mouth; and he was taken on board alive, for the naturalist to examine . . .

We were proceeding onward with our prize, when a white eagle, with fierce aspect and outspread wing, was seen bounding towards us; but stopping short, at twenty yards off, he flew up into a tree.

Another bird of the same kind discovered himself by making a motion to pounce down upon us as we passed underneath; and it seemed evident that they took us for kanguroos, having probably never before seen an upright animal in the island, of any other species. These birds sit watching in the trees, and should a kanguroo come out to feed in the day time, it is seized and torn to pieces by these voracious creatures . . .

White-breasted sea-eagle, *Haliaeetus leucogaster*

White-breasted sea-eagle, *Haliaeetus leucogaster*

Western grey kangaroo, *Macropus fuliginosus*

Matthew Flinders

When the ice-caps of the last Pleistocene ice-age slowly melted some 14 000 years ago, ocean levels round the world rose an estimated 70 to 80 metres. More than any other geological event, the resulting world-wide flooding of coastal lands shaped the continents as we now know them.

In our region, Tasmania once more became an island and New Guinea became separated from Australia by the creation of what we now call Torres Strait. Sydney Harbour was born when the valley of a small coastal stream was engulfed.

In what is now South Australia, an extension of the Flinders Ranges-Mt Lofty system was cut off by the rising waters of Backstairs Passage. It became Kangaroo Island.

Some of the earliest evidence of a human presence in Australia, artefacts some forty to fifty thousand years old, have been found on Kangaroo Island. But from the evidence of Flinders's eyes, those people must have been long gone by March 1802, when the *Investigator's* company became the first-known Europeans to set foot on this beautiful, tranquil island.

The first living things the *Investigator*'s company saw, even before they went ashore, were the kangaroos for which Flinders named the place. They became the nominate, or first described, race of the western grey kangaroo, *Macropus fuliginosus.*

Obviously unused to mankind, they behaved like perfect dodos and the details of their slaughter make unpleasant reading.

But then Flinders was meticulous about standard of diet on board, insisting on fresh food wherever it could be found.

That justification aside, the *Investigator*'s arrival at Kangaroo Island, as Flinders himself recognised, brought to a violent end its splendid isolation.

. . . the extraordinary tameness of the kanguroos

It was too late to go on shore this evening; but every glass in the ship was pointed there, to see what could be discovered.

Several black lumps, like rocks, were pretended to have been seen in motion by some of the young gentlemen, which caused the force of their imaginations to be much admired; next morning, however, on going toward the shore, a number of dark-brown kanguroos were seen feeding upon a grass plat by the side of the wood; and our landing gave them no disturbance.

I had with me a double-barrelled gun, fitted with a bayonet, and the gentlemen my companions had muskets. It would be difficult to guess how many kangaroos were seen; but I killed ten, and the rest of the party made up the number to thirty-one, taken on board in the course of the day; the least of them weighing sixty-nine, and the largest one hundred and twenty-five pounds.

These kanguroos had much resemblance to the large species found in the forest lands of New South Wales; except that their colour was darker, and they were not wholly destitute of fat.

After this butchery, for the poor animals suffered themselves . . . in some cases to be knocked on the head with sticks, I scrambled with difficulty through the brush wood, and over fallen trees, to reach the higher land with the surveying instruments . . .

There was little doubt that this extensive piece of land was separated from the continent; for the extraordinary tameness of the kanguroos and the presence of seals upon the shore, concurred with the absence of all traces of men to show that it was not inhabited.

The whole ship's company was employed this afternoon, in skinning and cleaning the kanguroos; and a delightful regale they afforded, after four months privation from almost any fresh provisions. Half a hundred weight of heads, fore quarters, and tails were stewed down into soup for dinner on this and the succeeding days; and as much steaks given, moreover, to both officers and men, as they could consume by day and by night. In gratitude for so seasonable supply, I named the southern land Kanguroo Island.

Matthew Flinders

Flinders postulated two possible causes for the evidence of fires he saw on uninhabited Kangaroo Island and islands in Spencer Gulf: they were caused either by lightning or by unknown earlier visiting seamen, perhaps even the long-missing French explorer La Perouse.

Given the evidence Flinders provides, there was perhaps no great mystery. Lightning was, and remains, one of the important causes of Australian bushfires.

On an uninhabited island where perhaps no Aboriginal fires had been lit for millennia, the woodland floor would soon become deep in leaf and branch litter. Under such conditions, in a prolonged dry summer, a lightning strike could cause violent wildfire, resulting in the very appearance Flinders describes.

The common seal on Kangaroo Island was then, as now, the Australian sea-lion, *Neophoca cinerea.* Contrasting the absence of timidity in the kangaroos, its comparative shyness and its speed of perception about men probably reflected its knowledge of Aboriginal hunters on nearby mainland beaches—knowledge safely denied the landlocked kangaroo until Europeans arrived.

Never perhaps had the dominion possessed here by the kanguroo been invaded before . . .

A thick wood covered almost all that part of the island visible from the ship; but the trees in a vegetating state were not equal in size to the generality of those lying on the ground, nor to the dead trees standing upright. Those on the ground were so abundant, that in ascending the higher land, a considerable part of the walk was made upon them.

They lay in all directions, and were nearly of the same size and in the same progress towards decay; from whence it would seem that they had not fallen from age, nor yet been thrown down in a gale of wind. Some general conflagration, and there were marks apparently of fire on many of them, is perhaps the sole cause which can be reasonably assigned; but whence came the woods on fire?

That there were no inhabitants upon the island, and that the natives of the continent did not visit it, was demonstrated, if not by the want of all signs of such visit, yet by the tameness of the kangaroo, an animal which, on the continent, resembles the wild deer in timidity . . .

. . . Never perhaps had the dominion possessed here by the kanguroo been invaded before this time. The seal shared with it upon the shores, but they seemed to dwell amicably together.

It not unfrequently happened, that the report of a gun fired at a kanguroo near the beach, brought out two or three bellowing seals from under bushes considerably further from the water side.

The seal, indeed, seemed to be much the most discerning animal of the two; for its actions bespoke a knowledge of our not being kanguroos, whereas the kanguroo not unfrequently appeared to consider us to be seals . . .

Australian sea-lions, *Neophoca cinerea*

Matthew Flinders

Australian pelicans, *Pelecanus conspicillatus*, are still common on Kangaroo Island. But Flinders was unhappily prophetic about the immediate future of those in Pelican Lagoon.

Sealers and whalers of many countries were soon to use the island as a base for shelter, water and supplies. American River, at the entrance of Pelican Lagoon, recalls the period.

Pelicans were fair game and are in any event very sensitive to disturbance at their breeding colonies, readily deserting eggs and young. So the name Pelican Lagoon truly recalls a Golden Age in natural Australia, brought suddenly to a close by the surging tide of European arrival.

The 'cassowaries' Flinders mentions were among the first victims of this tide. This entry (and a few lines written earlier, on 22 March 1802) record the discovery of the dwarf emu, *Dromaius minor*, on Kangaroo Island.

This bird differed specifically from the emu, *D. novaehollandiae*, of the mainland (and formerly Tasmania). It is now known from only a few museum specimens, of thousands slaughtered by seamen and sealers in the first few decades of last century on Kangaroo and King islands in Bass Strait.

The species was found nowhere else but on those two once protectively remote islands. But in the space of a few decades every living trace of them was wiped from the face of the earth. In such ways did we begin to rid a supremely beautiful, endlessly interesting natural world of its living diversity.

Alas for the pelicans! Their golden age is past . . .

April 4th, 1802. The entrance of the piece of water at the head of Nepean Bay, [Kangaroo Island, SA] is less than half a mile in width, and mostly shallow . . . Boats can go to the head of the southern branch only at high water; the east branch appeared to be accessible at all times; but as a lead and line were neglected to be put into the boat, I had no opportunity of sounding.

There are four small islands in the eastern branch; one of them is moderately high and woody, the others are grassy and lower; and upon two of these we found many young pelicans, unable to fly.

Flocks of the old birds were sitting upon the beaches of the lagoon, and it appeared that the islands were their breeding places; not only so, but from the number of skeletons and bones there scattered, it should seem that they had for ages been selected for the closing scene of their existence.

Certainly none more likely to be free from disturbance of every kind could have been chosen, than these islets in a hidden lagoon of an uninhabited island, situate upon an unknown coast near the antipodes of Europe; nor can any thing be more consonant to the feelings, if pelicans have any, than quietly to resign their breath, whilst surrounded by their progeny, and in the same spot where they first drew it . . .

Alas, for the pelicans! Their golden age is past; but it has much exceeded in duration that of man.

I named this piece of water *Pelican Lagoon*. It is also frequented by flocks of the pied shag, and by some ducks and gulls; and the shoals supplied us with a few oysters . . .

Not less than thirty emus or cassowaries were seen at different times; but it so happened that they were fired at only once, and that ineffectually.

Australian pelicans, *Pelecanus conspicillatus*

31

Tabular coral, Great Barrier Reef

Matthew Flinders

After refurbishing HMS *Investigator* at Port Jackson, Flinders sailed north on 22 July 1802 to chart the east and as much of the north coasts as possible before the summer monsoon.

Leaving the immediate eastern Australian coast at the end of September 1802, Flinders began a hazardous survey of the innumerable coral reefs east and north-east of present Mackay. He was urgently seeking a break in the outer barrier that would let him through into the Coral Sea and away northward.

The strain shows occasionally in his log. On 5 October he remarks: 'From the high breakers seen in the afternoon, . . . hopes were entertained of soon clearing the reefs; for by this time I was weary of them, not only from the danger to which the vessels were thereby exposed, but from fear of the contrary monsoon setting in upon the North Coast, before we should get into the Gulph of Carpentaria'.

But they would not escape until 20 October, by which time Flinders, in some desperation at the sailing performance of his companion vessel, the brig *Lady Nelson*, had ordered her to return to Port Jackson.

(It helps to keep reminding oneself, in all this, that they were not in motor-driven vessels, but sailing ships very much at the command of the stiff winds and pouring tide-races that hazard this region.)

The description Flinders gives of his first close view of coral when he landed at low tide on Saturday 9 October 1802, is one of the best early descriptions we have of a European starting to come to terms with this overwhelmingly beautiful, dangerous region. Note his use of the phrase 'a new creation'.

. . . wheat sheaves, mushrooms, stags horns

In the morning we steered E.N.E., with a light air from the southward; the brig was a-head, and at half past nine, made the signal for immediate danger; . . . I sent the master to sound past the brig; and on his finding deeper water we followed, drifting with the tide. At eleven he made the signal for being on a shoal, and we came to, in 35 fathoms, broken coral and sand; being surrounded by reefs, except to the westward from whence we had come . . .

In the afternoon, I went upon the reef with a party of the gentlemen; and the water being very clear round the edges, a new creation, as it was to us, but imitative of the old, was there presented to our view.

We had wheat sheaves, mushrooms, stags horns, cabbage leaves, and a variety of other forms, glowing under water with vivid tints of every shade betwixt green, purple, brown, and white; equalling in beauty and excelling in grandeur the most favourite parterre of the curious florist.

These were different species of coral and fungus, growing, as it were, out of the solid rock, and each had its peculiar form and shade of colouring; but whilst contemplating the richness of the scene, we could not long forget with what destruction it was pregnant . . .

James Grant

In 1800, the British Admiralty sent out to Australia under Lieutenant James Grant a shallow draught coastal exploring vessel, the 60-ton brig *Lady Nelson*.

She became the first ship known to have sailed successfully through Bass Strait, approaching from the west and naming Cape Otway and Cape Schanck among other features as she passed through.

Reaching Port Jackson in December 1800, Grant was later sent back to Bass Strait by Governor King to more closely investigate the deep indentation in the coast between those capes.

Instead, Grant chose to concentrate on the further exploration of Westernport, discovered three years earlier by George Bass.

Approaching the western entrance to Westernport on 21 March 1801, Grant became the first-known European to see the great colony of Australian fur seals, *Arctocephalus pusillus*, on what we now call Seal Rocks, off the south-western tip of Phillip Island.

Reports like Grant's and those of Bass about other colonies brought a rush of sealers of many nationalities to Bass Strait with appalling results.

According to a recent paper by R. M. Warneke of the Victorian Fisheries and Wildlife Division, and P. D. Shaughnessy of the CSIRO Institute for Biological Resources, Canberra, the harvest of skins of Australian fur seals could have totalled 200 000 by 1825, when the back of the herds was broken.

Although largely protected since the 1890s, the Bass Strait fur seals have never fully recovered from that orgy of shooting, bludgeoning, skinning and boiling down.

Warneke and Shaughnessy estimate that in the last thirty years, the Bass Strait population of the Australian fur seal has now stabilised at some 20 000 animals—well below its pre-sealing levels.

. . . others remained on the rocks, making a most disagreeable noise

At four P.M. of the 21st we had sight of the Island which forms the south head of Western Port, having the likeness of a snapper's head, or horseman's helmet . . . [falling] in a high clay bluff down to the water's edge.

The small islands lying off from it were covered with seals, numbers of which on our approach precipitated themselves into the sea, covering the passage, while others remained on the rocks making a very disagreeable noise, somewhat like the grunting of pigs.

They were of a large size, many of them being nearly equal to that of a bullock . . . accordingly I named these, Seal Islands.

I sent a boat ahead to sound the passage, and found between the Seal Islands and the South Head twelve, nine, six, five, and three and a half fathoms water . . .

This passage will shorten the distance when there is a leading wind, but standing round to the westward of Seal Islands there will be found sufficient room for any number of vessels to beat in.

Mr. Bass, when he visited this place in the whale boat, entered the port by the eastern passage, . . . coasting the western shore, from whence he made his remarks.

It is probable that these islands, lying so close to the opposite side of him . . . did not shew themselves to be detached from the southern side of the entrance, and this I judge because he makes no mention of them.

And I am the more inclined to this opinion, as no one has ever thought of looking here for seals, notwithstanding they may be found in great numbers, with an excellent harbour, affording good shelter for vessels employed in pursuit of them . . .

Australian fur seals, *Arctocephalus pusillus*

George Bennett

Born in Plymouth in 1804, George Bennett first visited Australia as a ship's surgeon in 1829, a year after graduating. He made another visit in 1832 and as a result published *Wanderings in New South Wales, Singapore and China* (London, 1834).

Apparently unable to keep away from the place, Bennett finally settled in Sydney in 1836 and went on to become one of the luminaries of the day: successful physician, member of the university medical faculty, founding secretary of the Australian Museum, correspondent of the Zoological Society, London, enquiring naturalist and friend of such greats as John Gould and Professor Richard Owen.

He was a capable writer and his second book, *Gatherings of a Naturalist in Australasia* (London, 1860), illustrated by George French Angas, is a much-quoted local classic of 19th-century Australian natural history. Bennett was perhaps best known for his investigation of the platypus. Less well-known are his comments about some other of our animals.

The following is his description of the beautiful fluffy or yellow-bellied glider, *Petaurus australis*, still widespread but declining with clearing in the wetter forests of eastern mainland Australia.

Few of today's Australians know this delicious creature despite some excellent writing on gliders by zoologists like David Fleay.

Bennett's description of it should be qualified: as well as nectar and insects it also enjoys the sugary sap of certain species of eucalypts, which it drinks by cutting deep lesions with its teeth in the bark of favourite trees. Over years, such trees become characteristically pitted with 'sap-wells'.

It . . . passes from one tree to another by flying leaps . . .

The Long-tailed Flying Opossum, or Flying Squirrel of the colonists . . . , is widely distributed in the forests and scrubs of New South Wales. It is also known as the Yellow-bellied Flying Phalanger.

Having received from the district near Broulee, south of Sydney, from a station on the Mooruya River, through the kindness of Mr. Henry Clarke, a young female of this species which had been captured alive in the scrubs, I availed myself of the opportunity of observing its habits in captivity, having before seen it only in a wild state . . .

It retires either between the forked branches or in the hollow cavities of the tree during the day to sleep, and at night passes from one tree to another by flying leaps, aided by its parachute-like membrane, descending to the ground only from unavoidable necessity, such as the trees being so far apart as to render it impossible to traverse the space by leaping.

When pursued, it takes to the highest branches, and springs from tree to tree with great rapidity, reminding me of monkeys I had seen in the forests of Singapore, which, when frightened, exhibit a similar degree of activity . . . It is surprising to see it jumping from branch to branch and tree to tree, in the clear and delightful atmosphere of a fine Australian moonlight night, with so extraordinary a degree of skill and rapidity. But I remarked that the flying leaps were invariably downwards, in an oblique direction; and that when desirous of ascending, the creature would climb rapidly. . .

In Australia the blacks capture them for food, and having prepared them by singeing the fur, cook them with the skins on, which gives the meat a more delicate and juicy flavour; but by the colonists they are valued only for their fur, which for delicacy and beauty, almost equals that of the Chinchilla . . .

It is evident, from the fondness of this animal for sweets, that when the *Eucalypti* are in flower, it subsists upon honey, which the blossoms yield in very large quantities (the honey is in such abundance as to afford subsistence to honey-eating parrots and other birds, as well as to these animals, and also to myriads of insects of various species). When these have disappeared, it lives upon the nuts and young foliage, and probably, as is usual with honey-feeding animals, also upon insects . . .

Yellow-bellied glider, *Petaurus australis*

Thomas Mitchell

Major Sir Thomas Mitchell, as he is best known in Australia, was born in Stirlingshire, Scotland in 1792, when the colony at Sydney Cove was four years old.

He arrived in the growing settlement in New South Wales in 1827 as deputy surveyor general, after a military career with Wellington's army in the Peninsula war, during which he became a remarkably skilled surveyor and draughtsman.

A complex man, who lived by military discipline, yet who frequently flouted the instructions of colonial governors, Mitchell was described in an official report as 'turbulent and self-indulgent'. And yet he was also immensely capable, energetic, artistic and seemingly immune to physical discomfort.

Mitchell made four major Australian explorations between 1831 and 1846. The following extracts from his *Three Expeditions into the Interior of Eastern Australia* (London, 1839) are only a fraction of his notes on wildlife, yet they are full of perceptive observations.

Returning to Sydney in February 1832 from his first expedition to the Gwydir River and the Darling, Mitchell, following his outward track, left the Namoi, crossed the Peel River and came down much the same route still used by the New England Highway through the Liverpool Range between Quirindi and Scone, along the Kingdon Ponds.

On this route, the party crossed what Mitchell, whose weird place-names were often rendered even more infuriatingly obscure by what one takes to have been a broad Scots' accent, named 'the plains of Mullaba . . . under Mt. Ydire'.

In this exalted location they met emus still ignorant of the fact that horses were not just fellow browsing animals, but often carried men who, unlike Mitchell on this occasion, could be dangerous.

. . . these birds might be thus approached on horseback

As we travelled across the plains, on which the young verdure, first offspring of the late rain, already began to shoot — four emus were observed quietly feeding at no great distance, apparently heedless of our party.

I approached them with my rifle, on a steady old horse, and found that this large quadruped, however strange a sight, did not in the least alarm those gigantic birds, even when I rode close up.

I alighted, levelled my rifle over the saddle and fired, but missed, as I presumed, for the bird merely performed a sort of pirouette, and then recommenced feeding with the others as before.

I had no means of reloading without returning to the party, but I was content with discovering that these birds might be thus approached on horseback — for in general the first appearance of men, although miles distant, puts them at once to their speed, which, on soft loose earth, perhaps surpasses that of a horse . . .

Emus, *Dromaius novaehollandiae*

38

Thomas Mitchell

On Mitchell's second expedition, he and his party of twenty-four men, seven ox-carts and a small herd of bullocks and horses, crossed Goobang Creek, west of the present towns of Parkes and Peak Hill, on 13 April 1835.

They moved out north-west into open forest country, into the watershed of the northward flowing Bogan River on their way to the Darling.

In June and July 1835, Mitchell travelled some 500 kilometres down the Darling from Fort Bourke, before returning to that depot without learning the river's final destination, as directed by Governor Bourke.

Whatever Mitchell's motives, and remembering his tragic clashes with the numerous Darling Aborigines, his descriptions of these people, healthy, proud of themselves and of their astounding hunting skills, make immensely sad reading.

Today when you confront an unpeopled, muddy river, its waters made turgid by European carp, its surroundings often eaten bare by sheep and cattle, you wonder could it have really been like this?

. . . they killed . . . some enormous cod-perch

These tribes inhabiting the banks of the Darling may be considered Icthyophagi [fish-eaters], in the strictest sense, and their mode of fishing was really an interesting sight.

There was an unusually deep and broad reach of the river opposite to our camp, and it appeared that they fished daily in different portions of it, in the following manner.

The king stood erect in his bark canoe, while nine young men, with short spears, went up the river, and as many down, until, at a signal from him, all dived into it, and returned towards him, alternately swimming and diving; transfixing the fish under water, and throwing them on the bank.

Others on the river brink speared the fish when thus enclosed, as they appeared among the weeds, in which small openings were purposely made that they might see them.

In this manner, they killed with astonishing despatch, some enormous cod-perch; but the largest were struck by the chief from his canoe, with a long barbed spear.

After a short time, the young men in the water were relieved by an equal number; and those which came out, shivering, the weather being very cold, warmed themselves in the centre of a circular fire, kept up by the gins on the bank . . .

Thomas Mitchell

Mitchell's picture of Aboriginal fishing activities was matched by detailed observations of the other habits and customs of these fascinating people: their hunting ways, the huts they built, their language, their dress, their ceremonies.

Mitchell's account of the manufacture of long nets—the Darling is up to 100 metres or more wide—and the way they were applied for catching waterfowl that probably included wood duck, *Chenonetta* *jubata;* black duck, *Anas superciliosa,* and grey teal, *A. gibberifrons,* is valuable and fascinating.

The account might seem far-fetched if it were not for two things: the skill of Aborigines in anything connected with hunting and the corroborating descriptions of the use of such nets by Eyre, Sturt, James Dawson and others.

That the Darling people used them as described, there is no doubt. This, again, is something now wholly gone.

. . . ducks . . . they ensnare with nets

The natives of the Darling live chiefly on the fish of the river, and are expert swimmers and divers . . . They also feed on birds, and especially on ducks, which they ensnare with nets, in the possession of every tribe.

These nets are very well worked, much resembling our own in structure, and they are made of the wild flax, which grows in tufts near the river.

These are easily gathered by the gins, who manage the whole process of net-making. They give each tuft (soon after gathering it) a twist, also biting it a little, and in that state it is laid about on the roof of their huts until dry.

Fishing nets are made of various similar materials, being often very large; and attached to some of them, I have seen half-inch cordage, which might have been mistaken for the production of a rope-walk.

But the largest of their nets are those set across the Darling for the purpose of catching the ducks which fly along the river in considerable flocks. These nets are strong, with wide meshes; and when occasion requires, they are stretched across the river from a lofty pole erected for the purpose on one side, to some large opposite tree on the other.

Such poles are permanently fixed, supported by substantial props . . . The native knows well "the alleys green" through which at twilight, the thirsty pigeons and parrots rush towards the water; and there, with a smaller net hung up, he sits down, and makes a fire ready to roast the birds, which may fall into his snare . . .

Wood duck, *Chenonetta jubata*

Thomas Mitchell

Concluding his account of the Darling River in the winter of 1835, Major Mitchell summed up some of the birds.

Today, while the splendid red-tailed black cockatoo, *Calyptorhynchus magnificus,* still extends south to the Darling, the sulphur-crested cockatoo, *Cacatua galerita*, said by Mitchell to be common, is now locally uncommon. However, the similarly white-plumaged little corella, *Cacatua sanguinea,* not mentioned in this passage by Mitchell, is now locally abundant. Could one have replaced the other?

There is no doubt about the continued presence on the Darling of the gorgeous pink cockatoo, *Cacatua leadbeateri*, also known as the Major Mitchell cockatoo, from Mitchell's glowing account of it. With its delicate sunset pink underwings it certainly seems too ethereal a creature for almost any region.

But not so; complementing the range of the sulphur-crested cockatoo and still fairly common *in suitable habitat* across the vast arid region, pink cockatoos are seldom far from trees: river gums, or patches of mallee, native pines, acacias, casuarinas, often in conjunction with spinifex, *Triodia,* and other sandhill vegetation. Its scarcity along the plains of the Darling then, as now, was due to the comparative absence of suitable habitat.

. . . this beautiful bird, might have embellished the air of a more voluptuous region

The bronze-wing pigeon was here, as elsewhere, the most numerous of that kind of bird. Next in abundance was the crested pigeon, which seems more peculiar to these low levels . . .

The large black cockatoo was sometimes seen, and about the river banks, the common white cockatoo with yellow top-knot . . . The smaller bird of this genus with a scarlet and yelow crest, and pink wings . . . was rarely noticed, and it appeared to come from a distance, flying usually very high.

The pink-coloured wings and glowing crest of this beautiful bird, might have embellished the air of a more voluptuous region; and, indeed, from its transient visits, it did not seem quite at home on the banks of the Darling . . .

Pink cockatoos, *Cacatua leadbeateri*

OPPOSITE: Pink cockatoo, *Cacatua leadbeateri*

Charles Darwin

In 1836, Charles Darwin, (1809–82), later to be co-author of the theory of evolution, called briefly at Sydney in HMS *Beagle* on the way 'home' from the famous cruise to the South America.

While at Sydney, he took opportunity to make a quick journey over the convict-made road over the Blue Mountains to Bathurst. It was during this interlude that he recorded these comments.

Darwin's view of the future of the emu and the kangaroo, in this case the eastern grey, *Macropus giganteus*, may have been too bleak in an Australia-wide sense, but locally they proved correct. His comment about the plight of the Aborigines and their land was precise and prophetic.

There is no certainty which of the several local species of small marsupials, called 'kangaroo-rats' by the settlers, Darwin saw so unceremoniously chased into a tree by the greyhounds.

I have elected to illustrate a rufous rat-kangaroo or bettong, *Aepyprymnus rufescens*. One of the smallest of the kangaroos, it was once common in south-eastern Australia, but is now extinct over the southern part of its former range and uncommon in the more heavily settled parts of the rest, up the east coast.

A few years since, this country abounded with wild animals . . .

Early on the next morning, Mr Archer . . . had the kindness to take me out Kangaroo-hunting. We continued riding the greater part of the day, but had very bad sport, not seeing a kangaroo, or even a wild dog. The greyhounds pursued a kangaroo rat into a hollow tree, out of which we dragged it: it is an animal as big as a rabbit, but with the figure of a kangaroo.

A few years since, this country abounded with wild animals; but now the emu is banished to a long distance, and the kangaroo is become scarce; to both, the English greyhound is utterly destructive. It may be long before these animals are altogether exterminated, but their doom is fixed.

The natives are always anxious to borrow the dogs from the farm-houses: the use of them, the offal when the animal is killed, and milk from the cows, are the peace-offerings of the settlers, who push further and further towards the interior. The thoughtless aboriginal, blinded by these trifling advantages, is delighted at the approach of the white man, who seems predestined to inherit the country of his children . . .

Rufous bettong, *Aepyprymnus rufescens*

Platypus, *Ornithorhynchus anatinus*

Charles Darwin

Charles Darwin, who recorded these comments on the platypus, *Ornithorhynchus anatinus*, while travelling from Sydney to Bathurst in 1836, had the advantage of seeing the creature alive and swimming and then of handling a freshly-killed one. Even then, he thought the reality amazing enough.

How much more amazing did English zoologists and anatomists find this ancient Australian, when, at first, all they had to work with were skins on which the soft rubbery bill and the swimming membranes on the feet had shrunk and hardened like black, sun-dried leather.

Some even formed the view that the whole thing was a fake, put together from parts of other animals by skilled Chinese taxidermists. So far was European credulity stretched in coming to terms with the living reality of this new land.

. . . certainly it is a most extraordinary animal

In the dusk of the evening I took a stroll along a chain of ponds, which in this dry country represented the course of a river, and had the good fortune to see several of the famous Platypus, . . .

They were diving and playing about the surface of the water, but showed so little of their bodies that they might easily have been mistaken for water-rats. Mr Browne shot one: certainly it is a most extraordinary animal; the stuffed specimens do not at all give a good idea of the recent appearance of its head and beak; the latter becoming hard and contracted . . .

Joseph Hawdon

Born at Walkenfield, Durham, in 1813 and coming to Australia in 1834, Joseph Hawdon was a bold and enterprising colonist. In 1836, he became one of the first to overland stock from Sydney to the newly-settled Port Phillip district.

In 1838, when Adelaide was anticipating famine, Joseph, with Charles Bonney and nine men overlanded 'between 300 and 400' cattle from Howlong to the hungry town.

They were the first overlanders to Adelaide and got a hero's welcome. Hawdon dined with Governor Hindmarsh and was accorded a public dinner. Those who missed the dinner 'displayed their sympathy in the general joy, by roasting a bullock whole, and regaling themselves in the true spirit of Good Old England', wrote Hawdon.

The course Hawdon followed ran mostly through northern Victoria, ascending Murray tributaries like the Goulburn and the Loddon for safer crossing. Near present Blanchetown, they left the river and struck south-west. Checked in the steep parts of the Mt Lofty Ranges, they safely reached Adelaide early in April 1838, less than three months after leaving Howlong.

Hawdon was a good observer and naturalist and a competent leader. He lost no men, kept friendly relations with the Aborigines and caused them no injury.

This excerpt from his journal comes as the party crosses the plains of northern Victoria, to climb Mt Hope near Pyramid Hill, on the way to the Loddon River, which Major Mitchell with his liking for peculiar names had called the 'Yarane' two years before.

The wildlife seen by Hawdon is of great interest. Then the plains of northern Victoria had perhaps a dozen species of small native mammals that are now locally or wholly extinct.

What Hawdon's 'small kangaroos' may have been, can only be guessed.

One was perhaps an eastern hare-wallaby, *Lagorchestes leporides*, once common on the Murray plains, now extinct as a result of European settlement.

The 'white macaws . . . with the long, hooked upper bills', can only have been long-billed corellas, *Cacatua tenuirostris*, although their heads are far from crimson. They later mostly disappeared from these northern plains but are now spreading north and east again after a resurgence in Victoria's Western District.

On 11 February 1838, Joseph Hawdon's party arrived at Swan Hill, previously named by Mitchell. Hawdon describes such a scene of teeming life and sylvan beauty on the lakes that one longs to have seen it.

Here, apart from 'hundreds' of brolgas, bustards and waterbirds, Hawdon met a beautiful parrot which seems from his description to have been the mallee ringneck, *Barnardius barnardi*.

These excerpts are from Hawdon's *The Journal of a Journey from New South Wales to Adelaide* (Georgian House, Melbourne, 1952).

. . . one of the most beautiful birds I ever beheld

February 3rd. At sunrise we started to cross the dry country towards the River Yarane. Our first eight miles were through a thick bushy scrub, full of a small description of Kangaroos and Emus; after which we again entered upon extensive plains.

We then passed a high rocky hill, and four miles further went close under "Mount Hope", also high and rocky, and of sugar-loaf shape. Mr. Bonney and I ascended to its summit, which terminated in a pinnacle of rocky granite. From this eminence we had a most extensive view . . . This hill appears to be inhabited by a small Kangaroo of a fawn colour, with a most beautiful head, and about the size of an English hare. I shot one, and preserved its skin . . .

White macaws with crimson heads fly about these plains in such numerous flocks that sometimes as they passed screaming over our heads they almost darkened the air. I shot a few of them. Their upper bill is hooked, and much larger than the lower one. Their maws were full of small bulbous roots, which they dig out of the plains with their bills. . .

February 11th. Being Sunday, I purposed resting, but as no wood could be found for our fires, we passed on twelve miles, and at a place where this branch of the Hume again joins the main channel we encamped under a small hill on our left, covered on its sides with clumps of pine.

This is the "Swan Hill" of Major Mitchell . . .

In the morning, passing through a large flat, we went between two lakes, called by Major Mitchell "Lake Boree". The south one is encircled by a grassy ridge . . .

This lake was covered with thousands of birds of various descriptions. On the water and by its margin were seen Emus, Native Companions . . . in hundreds, wild turkeys . . . Black Swans, Pelicans, and ducks of every variety, with great numbers of the rose-coloured parrots.

I shot some beautiful small parroquetts; their tail, breast and belly were yellow, their wings light blue, their back green, and around the fore-part of the head, immediately above the bill, was a circle of crimson; altogether forming one of the most beautiful birds I ever beheld . . .

OPPOSITE: Brolgas, *Grus rubicundus*

46

Charles Sturt

Captain Charles Sturt, (1795–1869), perhaps the most gentle and gentlemanly of all inland Australian explorers, was driven by a conviction, not at all soundly based, that Australia had an inland sea.

He had come to Australia from England in 1826 as military secretary to Governor Darling. Finding to his surprise that he rather liked Australia, in 1828–29, backed by Darling, Sturt began his exploring career by a fruitful expedition inland down the Bogan and Macquarie rivers, which terminated in his discovery of the Darling. The ultimate course of the Darling, whether south to join the Murray or west to some unknown fate, became a matter for speculation.

In November 1829, Darling despatched Sturt on a larger expedition to explore the Murrumbidgee. Thus began Sturt's courageous journey by open boat in which he discovered the Murray and followed it down through Lake Alexandrina to the sea.

From there, with provisions low and men exhausted, he had to row back upstream to their starting place, where they arrived, half dead, in March 1830. On the way, Sturt went blind from scurvy and exhaustion and never fully regained his eyesight.

These expeditions and Eyre's demonstration that no large rivers flowed south from the inland into the Great Australian Bight, worked in Sturt's mind.

He became depressed, feeling his exertions had produced little of community benefit. Deprived of support when Governor Darling returned to England, harassed by the contemptuous antagonism of Major Mitchell, who had become Surveyor General in New South Wales in 1828, Sturt moved to South Australia. There, occupying several government posts, he nursed the growing conviction that an inland sea existed.

This obsession led to Sturt's final great expedition of 1844–46. From Adelaide, he went up the Murray to the Darling, north to Depot Glen in the Barrier Range, then far north across Cooper's Creek into Sturt's Stony Desert.

Anyone who has travelled in far-western New South Wales will remark Sturt's astonishing luck in finding Depot Glen, north of Milparinka. But was it luck? Alert to the signs, he located it by noticing the direction followed by a group of Aborigines.

As usual, he directed the credit elsewhere. 'Providence had . . . guided us to the only spot, in that wide-spread desert, where our wants could have been permanently supplied'.

From afar the small gorge which Depot Creek cuts in the low range of slaty rock is hardly visible, save for a tell-tale line of river redgums marking the sandy track of the creek out onto the plain.

But in that gorge are a series of long pools of cool muddy water, deep enough and protected enough to withstand the worst droughts. Euros, as is their way, hop stolidly up the rocky hillslopes as you approach; peaceful doves call in the trees; black-fronted plovers dip at the water-edge.

Finding the depot may have been providential, but it led to 'ruinous detention', illness from scurvy for much of the party, and death from the same cause of Sturt's assistant, James Poole. In time, Sturt's 'siege' at Depot Glen was seen as one of the great feats of nerve and endurance in Australian exploration.

Walled in by drought, the party was trapped from 27 January to 17 July 1845. 'We were locked up in that desolate and heated region . . . ' said Sturt, 'as effectually as if we had wintered at the Pole'.

With no way of knowing the permanence of the water, their stay was made the more nerve-wracking by seeing its level slowly sink. And there were other terrifying signs, not lost on the observant Sturt.

At the end of his *Narrative of an Expedition into Central Australia* (London, 1849), Sturt discusses the animals observed or collected on the expedition, enabling those mentioned at Depot Glen to be identified with reasonable certainty.

The parrots were mostly cockatiels, *Nymphica hollandicus*; and probably blue bonnets, *Psephotus haematogaster*, and red-rumped parrots, *P. haematonotus*. The paroquets were budgerigars, *Melopsittacus undulatus*.

The bitterns Sturt mentions are a

Little corellas, *Cacatua sanguinea*

surprise: he described them as having 'a black body, and a white neck with a light shade of yellow, and speckled black'. They camped in trees at the Depot by day and flew out to neighbouring water at night. From this description there can be little doubt that the birds were black bitterns, *Dupetor flavicollis*, far from the nearest-recorded occurrence. They gathered at Depot Creek at the end of April, and later departed.

The cockatoos were mostly little corellas, *Cacatua sanguinea*, of which there was 'an immense flock' on the plains near the Depot; there were also galahs, *C. roseicapilla*. Both are still there in numbers today.

... we were soon wholly deserted

The three last days of February were cool in comparison to the few preceding ones. The wind was from the south, and blew so heavily that I anticipated rough weather . . .

But . . . all these favourable signs vanished, the thermometer ascending to more than 100°.

When we first pitched our tents at the Depot the neighbourhood of it teemed with animal life. The parrots and paroquets flew up and down the creeks collecting their scattered thousands, and making the air resound with their cries.

Pigeons congregated together; bitterns, cockatoos, and other birds; all collected round as preparatory to migrating. In attendance on these were . . . hawks of different kinds, making sad havoc amongst the smaller birds.

About the period of my return from the north they all took their departure, and we were soon wholly deserted. We no longer heard the discordant shriek of the parrots, or the hoarse croaking note of the bittern. They all passed away simultaneously in a single day; the line of migration being directly to the N.W. . . .

49

Charles Sturt

At last, in July 1845, Sturt's relief at Depot Glen came with a sudden change of weather. He wrote:

On the 11th the wind shifted to the east, the whole sky becoming overcast, at noon [of the 12th] it veered round to the north, when a gentle rain set in, so gentle that it resembled a mist, but this continued all the evening and during the night . . .

All night it poured down . . . and as morning dawned the ripple of waters . . . was a sweeter and more soothing sound than the softest melody I ever heard . . .

The moment of our liberation had arrived.

But it was too late for poor James Poole, who suddenly worsened and died from scurvy and malnutrition. They buried him under a grevillea tree which, now sparse and old, still stands there, with the initials, 'J.P. 1845', carved in its trunk.

Sending a party to the south with news of their position, and with their wagon wheels now sinking in the rain-soaked ground, on 18 July 1845, Sturt's party set off north-west, towards what he was to call Fort Grey in the far north-west corner of New South Wales.

Here, during the evening, they met a species of hopping-mouse, probably the dusky, *Notomys fuscus*, a beautiful, little-known native rodent which formerly ranged from north-west New South Wales across to the Nullarbor Plain.

Before European settlement brought feral cats to the inland it was immensely abundant. Elsewhere, Sturt wrote of Aborigines coming in from the sandhills 'with bags full', 150 to 200 at a time.

Writing of 'Mitchell's hopping mouse', but more likely meaning this species, Sturt remarked:

It was . . . found in vast numbers, inhabiting the sandy ridges from Fort Grey to Lake Torrens. Those immense banks of sand were in truth worked over with their footprints as if an army of mice or rats had been running over them . . . They are taken by the natives in hundreds, who avail themselves of a fall of rain to rove through the sandy ridges to hunt these little animals and the talpero [a bandicoot] as long as there shall be surface water . . .

The dusky hopping-mouse is now found, in much smaller numbers, only in pockets in north-east South Australia and far south-west Queensland. (Sturt uses the European name 'jerboa': the name 'hopping-mouse' for these little Australian rodents was not yet in vogue.)

. . . they hopped on their hind legs . . . like the kangaroo

Whilst we were here encamped a little jerboa was chased by the dogs into a hole close to the drays; which, with four others, we succeeded in capturing, by digging for them.

This beautiful little animal burrows in the ground like a mouse, but their habitations have several passages, leading straight, like the radii of a circle, to a common centre, to which a shaft is sunk from above, so that there is a complete circulation of air along the whole.

We fed our little captives on oats, on which they thrived, and became exceedingly tame. They generally huddled together in a corner of their box, but, when darting from one side to the other, they hopped on their hind legs, which, like the kangaroo, were much longer than the fore, and held the tail perfectly straight and horizontal. At this date they were a novelty to us, but we subsequently saw great numbers of them, and ascertained that the natives frequented the sandy ridges in order to procure them for food . . .

Dusky hopping-mouse, *Notomys fuscus*

Spinifex pigeon, *Petrophassa plumifera*

Charles Sturt

In his quest for an inland sea, Sturt pressed steadily north-west from his depot at Fort Grey until he reached—and named—Cooper's Creek. Turned back by what became known as Sturt's Stony Desert, he retreated to his depot before making one final attempt to discover his inland sea.

Sturt pushed the party steadily on over gibber deserts, sand-dunes and salty wastes, across the eastern Simpson Desert, across what we now call the Birdsville Track, across the present South Australian border, to a bearing near the present site of Birdsville in far south-west Queensland. Here, nearly 80 kilometres beyond their last watering-point, horses exhausted (one dropped dead a few hours later), with certain destruction ahead and rapidly drying waterholes behind, Sturt faced the decision of his life.

'. . . I lingered undecided on the hill, reluctant to make up my mind . . . I might, it is true, have gone on and perished with all my men; but I saw neither the credit nor the utility in such a measure . . .'

By great good fortune they regained the temporary safety of Cooper's Creek. Unbelievably, on 6 November 1845, now ill and weak from scurvy, Sturt moved into broken range country north of Cooper's Creek in far south-west Queensland to again attempt striking north, if a way could be found.

Fortunately for his whole party it could not. But there was one slight reward. While in the rocky country north of the Cooper, Sturt saw and collected a beautiful little pigeon, the spinifex or plumed pigeon, *Petrophassa plumifera*.

A stocky little bird with quail-like mannerisms, it was near the south-eastern corner of its range.

Sturt's description comes from his journal entry for 6 November 1845, and from the appendix to volume two of his *Narrative*.

. . . a new and beautiful little pigeon

I could see nothing . . . to cheer me in the view that presented itself.

To the northward was the valley . . . bounded all round by barren, stony hills, like that on which I stood . . . To the east the apparently interminable plains on which we had been, still met the horizon . . . Westward the outer line of hills continued backed by others . . . Satisfied that my horses had not the strength to cross such a country, . . . I turned back to Cooper's Creek, . . .

In riding amongst some rocky ground, we shot a new and beautiful little pigeon, with a long crest . . . Its locality was . . . confined to about thirty miles along the banks of that creek, and it was generally noticed perched on some rock fully exposed to the sun's rays, and evidently taking a pleasure in basking in the tremendous heat. It was very wild and took wing on hearing the least noise, but its flight was short and rapid like that of a quail, which bird it resembles in many of its habits.

In the afternoon this little pigeon was seen running in the grass on the creek side, and could hardly be distinguished from a quail. It never perched on the trees, but when it dropped after rising from the ground, could seldom be flushed again, but ran with such speed through the grass as to elude our search . . .

James Backhouse

James Backhouse (1794–1869), an English botanist and nurseryman turned Quaker, spent the years 1832–37 touring the Australian colonies with a companion, holding services with other members of the Society of Friends.

Backhouse was an acute observer and a pleasant writer. His book *A Narrative of a Visit to the Australian Colonies* (London, 1843) contains much descriptive comment about countryside and wildlife.

In 1837 Backhouse visited Melbourne in the party of Lady Franklin, celebrated wife of the Tasmanian governor, *en route* to Sydney. Melbourne was not yet two years old and the Yarra was still a pristine, lovely river, overhung with wattles. Bell miners, *Manorina melanophrys*, and eastern whipbirds, *Psophodes olivaceus*, tinked and cracked from the riverside forests and thickets and nankeen night herons, *Nycticorax caledonicus*, with handsome double white filaments falling from their crowns, camped in the riverside thickets.

Today night-herons still camp (but do not breed) in the re-constituted habitat of the Royal Botanic Gardens, next to the river. The nearest bell miner colony is a few kilometres upstream near Yarra Bend, but to find Melbourne's nearest whipbirds you must go a considerable distance further up-river or to the Dandenongs, more than 30 kilometres from the city.

Backhouse's other observations about local birds are of less value in not being specific about place. But the list is sufficient to show that Melbourne's environs were well-tenanted. Backhouse's 'Round-headed White Cockatoos' were probably long-billed corellas, *Cacatua tenuirostris*, which subsequently disappeared from the Melbourne region. Only in recent decades has there been a resurgence of the species in the Western District.

... the shrill whistle of the Coachman

The Yarra-yarra River is deep; but it is difficult to navigate, for boats, on account of the quantity of sunken timber.

It is about sixty feet wide, and margined with trees and shrubs. Among these are heard, the tinkling note of the Bell-bird, and the shrill whistle of the Coachman, which is terminated by a jerking sound, something like the crack of a whip. We also noticed the Nankin-bird, a species of Heron, which is cinnamon coloured on the back, sulphur-coloured on the breast, and has a long, white feather, pendant from the back of the head.

The river is fresh to Melbourne, where there is a rapid. The country on its banks is open, grassy forest, rising into low hills. The town of Melbourne, though scarcely more than fifteen months old, consists of about a hundred houses, among which are stores, inns, a jail, a barrack, and a school-house. . . . The water continues fresh for a considerable distance below the town. Its banks are low, and fringed with bushes.

Toward the mouth of the river, there are swamps, covered with a narrow-leaved, white-flowered Melaleuca, drawn up like hop-poles, to thirty feet in height . . . Contiguous to the river, there are some beautiful pieces of land, clear of trees and covered with green grass. . . .

The emus are fast retiring before the white population, and their flocks and herds. The large bird of the crane kind, called here, the Native Companion, and a Bustard, denominated the Wild Turkey, are plentiful; there are also yellow-tailed Black Cockatoos, Round-headed White Cockatoos, Parrots of various kinds, Pelicans, Black Swans, Ducks, White Hawks, Laughing Jackasses, Kingfishers, Quails and various birds, not to omit the Piping Crow, with its cheerful note and the Black Magpie.

Nankeen night heron. *Nycticorax caledonicus*

John Gilbert

John Gilbert was an outstanding figure in 19th-century Australian zoological discovery. Before his arrival in Australia with the ornithologist John Gould in 1838, he had been a taxidermist with the Zoological Society of London.

Travelling widely, he went far beyond the fringe of settlement, where there was danger, but the wildlife was plentiful and undisturbed. Between 1838 and his violent death with Leichhardt's party on Cape York in June 1845, Gilbert sent back to Gould the finest collection of Australian fauna then in existence.

Coming from a country where birds make nests as birds should, he was naturally incredulous at the incubation mounds of the Australian birds known as megapodes, particularly of the scrubfowl, *Megapodius reinwardt*, whose acquaintance he made in mid-1840 near Port Essington on Cobourg Peninsula, on the far north-west tip of Arnhem Land.

Most megapodes live in the tropics of the Australian region, and most lay their eggs in huge heaps of damp earth mixed with sticks and leaf-debris, where oxidisation provides the necessary steady warmth for incubation.

The scrubfowl, described here by Gilbert, is remarkable for its enormous mounds. Two or three pairs may share the largest of these and keep them going for years. In these mounds the females lay their eggs at the bottom of deep, sloping shafts, which are then filled with loose material. The depth of these shafts presumably ensures good incubation and helps defeat raiding

For while the strong-legged bird (megapode means 'great-footed') handles the task with apparent ease, to a man it is not easy work digging out the packed, stick-filled earth in the steam-heat of a tropical day.

56

Scrubfowl, *Megapodius reinwardt*

. . . I was satisfied that these mounds had some connexion with the bird's . . . incubation

On my arrival at Port Essington my attention was attracted to numerous immense mounds of earth, which were pointed out to me by some of the residents as the tumuli of the aborigines . . .

On the other hand, I was assured by the natives that they were formed by the Megapode for the purpose of incubating its eggs: their statement appeared so extraordinary, and so much at variance with the general habits of birds, that no one in the settlement believed them or took sufficient interest in the matter to examine the mounds, and thus to verify or refute their accounts . . .

Another circumstance which induced a doubt of their veracity was the great size of the eggs brought in by the natives as those of this bird.

Aware that the eggs of *Leipoa* were hatched in a similar manner, my attention was immediately arrested by these accounts, and I at once determined to ascertain all I possibly could respecting so singular a feature in the bird's economy . . .

[Having] procured the assistance of a very intelligent native, who undertook to guide me to the different places resorted to by the bird, I proceeded on the sixteenth of November to Knocker's Bay, a part of Port Essington Harbour comparatively but little known, and where I had been informed a number of these birds were always to be seen.

I landed beside a thicket, and had not proceeded far from the shore ere I came to a mound of sand and shells, with a slight mixture of black soil, the base resting on a sandy beach, only a few feet above high-water mark; it was enveloped in the large yellow-blossomed *Hibiscus*, was of a conical form, twenty feet in circumference at the base, and about five feet in height. On pointing it out to the native and asking him what it was, he replied 'Oooregoorga Rambal,'—Megapode's house or nest.

I then scrambled up the sides of it, and to my extreme delight found a young bird in a hole about two feet deep; it was lying on a few dry withered leaves, and appeared to be only a few days old.

So far I was satisfied that these mounds had some connexion with the bird's mode of incubation; but I was still sceptical as to the probability of these young birds ascending from so great a depth as the natives represented; and my suspicions were confirmed by my being unable to induce the native, in this instance, to search for the eggs, his excuse being that 'he knew it would be useless, as he saw no traces of the old birds having recently been there.' . . .

I continued to receive the eggs without having an opportunity of seeing them taken from the mound until the 6th of February, when on again visiting Knocker's Bay I had the gratification of seeing two taken from a depth of six feet, in one of the larges mounds I had then seen.

In this instance the holes ran down in an oblique direction from the centre towards the outer slope of the hillock, so that, although the eggs were six feet deep from the summit, they were only two or three feet from the side. The birds are said to lay but a single egg in each hole, and after the egg is deposited the earth is immediately thrown down lightly until the hole is filled up; the upper part of the mound is then smoothed and rounded over . . .

John Gould

John Gould (1804–81) was probably the most prodigious entrepreneur in the whole field of natural history, of any age. Among lesser works and a host of scientific papers, he created nearly twenty huge, extraordinarily beautiful imperial folio works on the birds of a dozen countries.

I mean 'created' in the fullest sense: he did rough but lively sketches to guide the artists who designed the original paintings. He superintended the wet-stone lithography by which the final pencilled design of those plates was transferred to paper. He supervised the subsequent hand-colouring of those lithographed pages by a team of artists.

Gould's wife Elizabeth, a graceful and astonishingly productive artist and young mother of six, painted (and in many cases transferred to stone) the original designs for the first great works that made Gould's name, including some for *The Birds of Australia*. In all, she painted some 600 of the 3000 plates in all the great books. After her death in 1841 at the age of 37, other artists took on the exacting work.

Each many-volumed work was organised, laid out in scientific order and the graceful text written by Gould: *A Century of Birds from the Himalaya Mountains, The Birds of Europe, The Birds of Australia* and supplements, *The Mammals of Australia, A Monograph of the . . . family of Humming-birds, The Birds of Asia, The Birds of Great Britain, The Birds of New Guinea* — and so the titles roll.

As well, Gould described hundreds of new species and maintained correspondence with scientists and collectors round the world.

Gould spent from September 1838 to April 1840 in Australia in order to more accurately describe the birds of this country. His suite included Elizabeth, then pregnant with their sixth child, and John Gilbert, his famous field-man, who was to

Little penguin, *Eudyptula minor*

58

contribute magnificently to Gould's Australian bird-collections.

Gould visited and collected in Tasmania and Bass Strait, South Australia and New South Wales. The result of all this, *The Birds of Australia*, of 600 plates, was issued in thirty-six parts between 1840 and 1848, with later supplements. Its subscribed price was £115. According to Gould, of the 250 sets printed, 238 subscriptions had been sold by 1866. A good set may now be worth $100 000.

The Mammals of Australia, in thirteen parts with 182 plates, was issued between 1845 and 1863. Never as keenly sought as the *Birds*, it is a handsome work with much original information.

The following excerpts come from those two great works, and from the smaller *Handbook of the Birds of Australia* (1865): milestones in wildlife exploration. Of all the books printed last century, perhaps none did more to make natural Australia known to Europeans than these.

The little penguin, *Eudyptula minor*, of southern Australian and New Zealand coasts, lives in temperate to warm waters where huge bulk is unnecessary to keep warm and where high-speed dashing about after fish would soon overheat, say, a huge emperor penguin.

But the little penguin pays a price for its smallness. It is easy prey to sea-eagles and harriers, at least when young. It was probably vulnerable to land-predators like the thylacine and later, the dingo. It is certainly vulnerable to the introduced fox in places like Victoria's Phillip Island.

So in Australia the little penguin is condemned to breed mostly on islands off the coast, to lay and brood its eggs in burrows and to come and go under the mantle and coolness of darkness. (Breeding in burrows and nocturnal land activity may also overcome an overheating problem for this still well-insulated bird during hot weather.)

Little penguin, *Eudyptula minor*

. . . it stems the waves of the most turbulent seas

This species is very abundant all round Tasmania, in Bass's Straits, and on the south coast of Australia generally, where it frequents those parts of the sea that are favourable to its habits and mode of life, and where the depth of the water is not too great to prevent its diving to the bottom.

It is also often seen in the deep bays and harbours, and some distance up the great rivers, but never I believe in fresh water; seas abounding in small islands, whose sides are not too precipitous for it to ascend for the purpose of breeding, being the localities most frequently resorted to. . . .

Its swimming powers are . . . so great, that it stems the waves of the most turbulent seas with the utmost facility, and during the severest gale descends to the bottom, where, among . . . forests of sea-weed, it paddles about in search of crustaceans, small fish, and marine vegetables, all of which . . . were found in the stomachs of those I dissected.

A considerable portion of the year is occupied in the process of breeding and rearing the young, in consequence of its being necessary that their progeny should acquire sufficient vigour to resist the raging of that element on which they are destined to dwell . . . Notwithstanding this care for the preservation of the young, heavy gales of wind destroy them in great numbers, hundreds being occasionally found dead on the beach after a storm; . . .

Some of the islands in Bass's Straits, where the Penguins are numerous, are completely intersected by paths and avenues, and so much care is expended by the birds in the formation of these little walks that every stick and stone is removed, and in some instances even the herbage, by which the surface is rendered so neat and smooth as to appear more like the work of the human hand than the labour of one of the lower animals.

Lesser noddies, *Anous tenuirostris*

John Gould

A relative of the terns, the lesser noddy, *Anous tenuirostris*, is a delicate brown seabird with a white cap. Found only in the Indian Ocean, its sole Australian breeding stations are on the Abrolhos Islands, off Geraldton, Western Australia.

Last century, its numbers were estimated in hundreds of thousands. In recent decades, several colonies (on Pelsart, Wooded and Morley islands in the Abrolhos) have been reckoned to total tens of thousands. The extent of any diminution is hard to demonstrate, but diminution there seems to have been.

Apart from visits by fishermen and tourists to these coral-girt islands, the main activity likely to have caused disturbance to the birds was a guano-gathering industry established in the mid-19th century. Over the years, workers took many birds and eggs for food, and on some of the islands, at least, they unintentionally left rats and mice.

John Gilbert visited the Abrolhos in the summer of 1842–43 and sent his employer John Gould the notes quoted here, from *The Birds of Australia.*

The Abrolhos colonies are still fascinating places. The dainty little terns come and go past your very face or sit watching you trustfully from a nest only a metre or two away, while the air is full of the gentle staccato rattle of their calls.

The birds seem to stay in the region of these islands all year, living on small fish taken from the surface of the sea by swooping, dipping flocks—noddies do not plunge. They gather the wet seaweed for their nests the same way, picking it from the surface of the sea in flight.

. . . their immense numbers strike you with astonishment

On the Houtmann's Abrolhos [the lesser noddy] is . . . numerous . . . [and] truly gregarious.

[The] nests [are] arranged as closely as possible on the branches of the mangroves, at a height of from four to ten feet above the ground, the sea-weed of which each nest is constructed being merely thrown across the branch, without any regard to form, until it has accumulated to a mass varying from two to four inches in height . . .

[In] many instances long pieces of sea-weed hang down beneath the branch, giving it the appearance of a much larger structure than reality; the nests and the branches of the trees are completely whitened with the excrement of the bird, the disagreeable and sickly odour of which is perceptible at a considerable distance.

South Island, Houtmann's Abrolhos, appears to be the only one resorted to for the purpose of nidification; for although large mangroves occur on others of the neighbouring islands, it was not observed on any of them.

"I have seen many vast flocks of birds", says Mr. Gilbert, "but I confess I was not at all prepared for the surprise I experienced in witnessing the amazing clouds, literally speaking, of these birds when congregating in the evening while they had their young to feed.

"Their alternate departure and return with food during the day, in the same route, had a most singular appearance.

"From their breeding-place to the outer reef, beyond the smooth water, the distance is four miles; and over this space the numbers constantly passing were in such close array that they formed one continuous and unbroken line.

"After the young birds were able to accompany their parents, I observed that they all left the breeding or roosting-place in the morning and did not again return until evening, the first-comers apparently awaiting the arrival of the last before finally roosting for the night.

"It is when thus assembling that their immense numbers strike you with astonishment. Even those who have witnessed the vast flights of the Passenger Pigeon, so vividly described by Audubon, could hardly avoid expressing surprise at seeing the multitudes of these birds, the quack and the piping whistle of the young ones are almost deafening . . . " ·

John Gould

Although he called the great egret, *Egretta alba*, 'The Australian Egret', John Gould knew that nearly identical races of this largest egret ranged through much of the rest of the world.

Now rare in much of Europe because of earlier shooting for the millinery plume trade, drainage of habitats and by hunting, it is still widespread in the Americas, Africa and Asia.

In Australia, despite slaughter of plume-bearing nesting adults on the Murray swamps and elsewhere at the turn of the century, the sight of a great egret wading is still surprisingly common.

You see them on tropical flood-plains and swamps, river-margins, tidal inlets and mudflats, drainage ditches and irrigation areas almost anywhere there are fish and frogs and a clear view all round. For great egrets, as Gould noted, are shy and suspicious.

Exceptions can be found at fishing places on coasts, estuaries, rivers and lakes where egrets, attracted to fish scraps, often become surprisingly tame. So the century and a half since Gould's visit has seen unpredictable developments in the fortunes of this handsome creature, most of them happier than he probably anticipated.

. . . its snowy plumage presents a . . . pleasing contrast

This noble . . . Egret, the largest of the group inhabiting Australia, is sparingly dispersed over all parts of that continent, and is usually met with along the rivers and lagoons of the interior as well as in the neighbourhood of the coast.

I have occasionally seen it near the mouth of the Hunter, but more frequently on the banks of the Clarence and other rivers little frequented by civilized man. I also observed it in Tasmania, in the vicinity of George's River, and the other unfrequented streams to the northward of the island. . . .

It is of an extremely shy and distrustful disposition, and can only be approached within range by the exercise of the utmost care and caution. Its powers of wing are considerable, and, like other Herons, it occasionally performs long-continued flights at a great height in the air; its food is also of a similar character, consisting of fish, frogs, aquatic insects, &c. When on the ground its snowy plumage presents a strong and pleasing contrast to the green sedge and other herbage clothing the banks of the rivers.

That it undergoes seasonal changes of plumage is evident, since I possess specimens, some of which are adorned with long ornamental plumes on the back, while in others they are entirely wanting, from which I infer that, as they all appear to be old birds, they have been killed at different periods of the year, and that these ornamental plumes are only carried during the months of spring and the breeding-season.

Great egret, *Egretta alba*

John Gould

Conscious as he was of the scientific importance of the birds he described, John Gould was also moved by beauty and was not hesitant to sketch a graceful word-picture of a bird that appealed to him.

It is interesting that he should have chosen to emphasise the then exclusively inland range of the attractive crested pigeon, *Ocyphaps lophotes*. He was not to know the bird's extreme adaptability or the extent of changes to inland habitats that were beginning.

Clearing of forests and woodlands, their replacement with openly-treed grasslands and in some cases cereal crops, the spread of introduced weeds and the provision of dams for stock-water over vast areas of Australia, all worked in the crested pigeon's favour, even as they dispossessed other wildlife.

From being restricted to inland plains, since Gould's day the bird has steadily worked its way coastwards in all mainland States. Happily one no longer needs to be a Sturt or a Leichhardt to see it.

The crested pigeon is opportunistic. It eats the seeds of such introduced weeds as Paterson's curse, *Echium lycopsis*, gleans grain from stubble and along roadsides and exploits food put out for poultry. In short, it is an adaptable, very successful and very beautiful bird.

Our only ground-dwelling grey pigeon with a crest, it can also be quickly identified in flight: bursts of whistling wingbeats alternate with fast, tilting glides on flat wings.

That characteristic wing-whistle, which comes from the third outer primary feather being narrowed toward the tip, inspires the country schoolchildren's name 'wirewings'. Possibly the sound, as one bird rapidly departs, alerts others to danger.

Crested pigeon, *Ocyphaps lophotes*

The chasteness of its colouring, the extreme elegance of its form . . .

The chasteness of its colouring, the extreme elegance of its form, and the graceful crest which flows backwards from its occiput, all tend to render this Pigeon one of the most lovely of its tribe inhabiting Australia; and in fact I consider it not surpassed in beauty by any other from any part of the world.

It is to be regretted that, owing to its being exclusively an inhabitant of the plains of the interior, it can never become an object of general observation; but . . . can only be seen by those of our enterprising countrymen whose love of exploring new countries prompts them to leave for a time the haunts of civilised man, to wander among the wilds of the distant interior . . .

It frequently assembles in very large flocks, and when it visits the lagoons or river-sides for water, during the dry seasons, generally selects a single tree, or even a particular branch, on which to congregate; very great numbers perching side by side, and all descending simultaneously to drink: so closely are they packed while thus engaged, that I have heard dozens of them being killed by the single discharge of a gun.

Its powers of flight are so rapid as to be unequalled by those of any member of the group to which it belongs; an impetus being acquired by a few quick flaps of the wings; it goes skimming off apparently without any further movement of the pinions. Upon alighting on a branch, it elevates its tail and throws back its head, so as to bring them nearly together, at the same time erecting its crest and showing itself off to the utmost advantage . . .

John Gould

In a sense, John Gould had no right to see budgerigars on the Mokai [Mooki] River, near Breza just west of the Divide in New South Wales. They seldom appear there in such numbers, but that year they came even further east.

As Gould remarked in a letter to E. P. Ramsay, ornithologist at the Australian Museum, Sydney: 'The Black fellows of the Upper Hunter told me that the little *Melopsittacus undulatu[s]* had come to meet me, for they had never seen the bird in the district until the year I arrived . . .'

Budgerigars are highly nomadic in response to inland rainfall, or the lack of it. They also make regular southward and eastward movements to breed in spring in areas of higher rainfall nearer the coasts.

In drought years, these movements often bring them further coastwards and in greater numbers. The likelihood that this happened in 1839 is strengthed by the fact that on the same expedition Gould collected the first specimen of a highly nomadic inland bird, the crimson chat, near the Peel River, east of the Liverpool Range, on 11 December 1839, decidedly out of its normal range.

Since Gould's day the picture has altered somewhat. Clearing and the establishment of countless stock-dams and tanks throughout the pastoral regions have favoured budgerigars. It is possible that they are now more widespread than ever. Gould's meeting with our now well-loved little green parrot had several results. He was able to sketch it from life, and Elizabeth made a charming plate of it.

Gould also brought young birds back with him to Sydney. 'I believe I was one of the first who introduced living examples to this country', Gould wrote in his *Handbook*. Those introductions eventually made our budgerigar the most phenomenally popular cage-bird ever.

. . . I saw them in flocks of many hundreds

I found myself surrounded by numbers, breeding in all the hollow spouts of the large *Eucalypti* bordering the Mokai; and on crossing the plains between that river and the Peel, in the direction of the Turi Mountain, I saw them in flocks of many hundreds, feeding upon the grass-seeds that were there abundant.

So numerous were they, that I determined to encamp on the spot, in order to observe their habits and procure specimens. The nature of their food, and the excessive heat of these plains, compel them frequently to seek the water; hence my camp, which was pitched near some small pools, was constantly surrounded by large numbers, arriving in flocks varying from twenty to a hundred or more.

The hours at which they were most numerous were early in the morning, and some time before dusk in the evening. Before going down to drink, they alight on the neighbouring trees, settling together in clusters, sometimes on the dead branches, and at others on the drooping boughs of the *Eucalypti*.

Their flight is remarkably straight and rapid, and is generally accompanied by a screeching noise. During the heat of the day, when flocks of them are sitting motionless among the leaves of the gum-trees, they are with difficulty detected.

Budgerigars, *Melopsittacus undulatus*

John Gould

Gould was fascinated by Australia's lorikeets: small, fast-flying, glossy-plumaged parrots whose brush-tipped tongues, specialised digestive systems and nomadic habits evolved to suit their diet of nectar and pollen of eucalypts, banksias and other flowering trees and shrubs.

The Hunter River region is home to four species of these vivacious parrots, whose astonishing energy is typical of birds with a high carbohydrate diet.

To see all four feeding in one enormous flowering eucalypt, and to collect all of them from the foliage of that one tree is an experience that would long live in the mind of any ornithologist, particularly one still coming to know this continent.

To me, this excerpt from Gould's *Handbook to the Birds of Australia* (1865), remains one of his most memorable passages. Happily, it is an experience that in somewhat lesser degree can still be shared by today's binocular and camera-carrying birdwatchers.

The species Gould saw were: the rainbow, scaly-breasted, musk and little lorikeets, respectively *Trichoglossus haematodus, T. chlorolepidotus, Glossopsitta concinna* and *G. pusilla.*

The incessant din produced by their thousand voices...

However graphically it might be described, I scarcely believe it possible to convey an idea of the appearance of a forest of flowering gums tenanted by *Trichoglossi*, three or four species being frequently seen on the same tree, and often simultaneously attacking the pendant blossoms of the same branch. The incessant din produced by their thousand voices, and the screaming notes they emit when a flock of either species simultaneously leave the trees for some other part of the forest, is not easily described, and must be seen and heard to be fully comprehended.

So intent are they for some time after sunrise upon extracting their honey-food, that they are not easily alarmed or made to quit the trees upon which they are feeding. The report of a gun discharged immediately beneath them has no other effect than to elicit an extra scream, or cause them to move to a neighbouring branch, where they again recommence feeding with avidity, creeping among the leaves and clinging beneath the branches in every variety of position.

During one of my morning rambles in the brushes of the Hunter, I came suddenly upon an immense *Eucalyptus*, which was at least two hundred feet high. The blossoms of this noble tree had attracted hundreds of birds, both Parrots and Honey-suckers; and from a single branch I killed the four species of the former inhabiting the district, viz. [the Rainbow, Scaly-breasted, Musk and Little Lorikeets.]

I mention this fact in proof of the perfect harmony existing between these species while feeding; a night's rest, however, and the taming effect of hunger doubtless contributed much to this harmonious feeling, as I observed that at other periods of the day they were not so friendly ...

Musk lorikeet, *Glossopsitta concinna*

John Gould

In *The Mammals of Australia* (London, 1845–1863), John Gould described the red kangaroo, *Macropus rufus*, as the finest of our indigenous animals.

It is able to survive in the harshest of environments, locating as if by magic each meagre pick of green food in depressions in the plains ,and in small drainage lines on the hills and slopes.

In so many ways, the animal is a match for its environment. When feeding is finished for the morning, as the inland day warms up, it digs itself a hip-hole in the shade of a shrub or tree. There, keeping cool and wasting as little body-moisture as possible, it lies quietly shaded during the heat of the day, nostrils and eyes narrowed to slits to reduce evaporation and glare and to defeat the attentions of small flies.

Gould's comments suggest that in 1839 the red kangaroo ranged somewhat further east in New South Wales than today. They also suggest that it had somewhat different habitat preferences before the spread of stock and clearing converted enormous areas of the inland from scrubland to grassland and provided innumerable dams and tanks.

That these changes, coupled with the decline of hunting by Aborigines, led to increases in its numbers seems undoubted.

Contrary to Gould, except in certain areas favoured by good seasonal conditions, the early inland explorers remarked on the scarcity of kangaroos. They saw and shot comparatively few. Most of them attributed this scarcity to the continual hunting pressure exerted by the Aborigines.

Taken collectively, these explorers' observations suggest that the red kangaroo was not a particularly abundant animal except where numbers congregated from a wide surrounding radius on green feed where good rains had fallen.

Red kangaroos, *Macropus rufus*

... the finest of the indigenous Mammals

Not only is this species the most beautiful member of the family to which it belongs, but it may also be regarded as the finest of the indigenous Mammals of Australia yet discovered; its large size, great elegance of form, and rich and conspicuous colouring all tending to warrant such an opinion . . .

The range of the Great Red Kangaroo, so far as it is yet known, extends over the plains of the interior of the Colonies of New South Wales, Port Philip, and South Australia . . . the plains bordering the rivers Gwydyr, Namoi, Morumbidgee, Darling and Murray, and the grassy hills of South Australia, particularly those to the northward of Adelaide, are the districts over which it formerly ranged in abundance, and in which, notwithstanding the persecution to which it has been subjected, it may still be found, though in much smaller numbers.

It does not so strictly affect the rich grassy plains as the [Eastern Grey] Kangaroo . . . but evinces a greater partiality for the side of the low stony hills and patches of hard ground clothed with box, intersecting those alluvial flats . . .

The female is still more gracefully and elegantly formed than the male, and has a very different style of colouring, delicate blue being the prevailing tint in those parts which in the male are red, whence the colonial names for the two sexes of Red Buck and Blue Doe; the female has also been called the Flying Doe, from her extraordinary fleetness, which is in fact so great, that I have no hesitation in saying that on hard ground, and under favourable circumstances, she would out-strip the fastest dogs . . .

John Gould

The thylacine, or Tasmanian 'tiger', *Thylacinus cynocephalus*, was by far the most formidable marsupial known to modern humans.

The largest thylacines shot or captured measured nearly two metres from nose to tail-tip. The strength of the huge jaws in the large, broad head was enormous.

So the thylacine was almost pre-destined to become a problem in Tasmania as sheep replaced kangaroos in the grasslands and open woodlands.

Trappers poisoned thylacines to reduce raids on possums and kangaroos snared for their dense winter fur. Pastoralists offered bounties on them and in the two decades after 1888, the Tasmanian government actually paid bounties on more than 2000 scalps.

The last officially-known thylacine was captured in 1933 and died in Hobart Zoo in 1936. Many supposed sightings have been made in Tasmania since, including an apparently authentic one by a National Park ranger in 1984.

But the animal's status hovers between 'excessively rare' and 'presumed extinct'—a victim of European thoughtlessness. It was also possibly victim of an epidemic disease that decimated marsupials about the turn of the century.

One thing John Gould wrote deserves comment. The thylacine did once inhabit mainland Australia and New Guinea. The fossil record reveals that it was widespread up to some 2000–3000 years ago.

The 'most recent' periods known of mainland occurrence overlap the Aboriginal presence by tens of thousands of years, and coincide with the arrival of the dingo in Australia 3000 to 4000 years ago.

Presumably the dingo helped wipe out the thylacine on the mainland but was prevented from doing so in Tasmania by the waters of Bass Strait, which last isolated the island State some 14 000 years ago.

It must be . . . the most formidable . . . of the indigenous mammals . . .

Tasmania, better known as Van Diemen's Land, is the country it inhabits, and so strictly is it confined to that island, that I believe no instance is on record of its having been found on the neighbouring continent of Australia.

It must be regarded as the most formidable, both of the Marsupialia and of the indigenous mammals of Australia; for although too feeble to make a successful attack on man, it commits sad havoc among the smaller quadrupeds of the country, and among the poultry, and other domestic animals of the settler; even sheep are not secure from its attacks, which are the more difficult to be guarded against, as the habits of the animal [are] nocturnal . . .

The destruction it deals around has, as a matter of course, called forth the enmity of the settler, and hence in all cultivated districts the animal is nearly extirpated; on the other hand, so much of Tasmania still remains in a state of nature, and so much of its forest land uncleared, that an abundance of covert still remains in which the animal is secure from the attacks of man; many years must therefore elapse before it can become entirely extinct; in these remote districts it preys upon [pademelons, Bennett's wallabies, bandicoots, echidnas] and all the smaller animals . . .

Mr. Ronald C. Gunn, who has had better opportunities than any other scientific man of observing the animal in its native wilds, states that it is common in the more remote parts of the colony, and that it is often caught at Woolnorth and the Hampshire Hills. He has seen some so large and powerful, that a number of dogs would not face one of them . . .

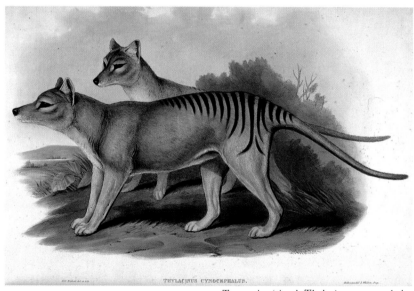

THYLACINUS CYNOCEPHALUS.

Tasmanian 'tiger', *Thylacinus cynocephalus*

John Gould

Anyone who has had the luck to see a honey possum, *Tarsipes spenserae*, in the wild has had a charming Australian experience.

John Gilbert clearly thought so when he communicated his observations of the little animal in southern Western Australia (its only home) to John Gould in 1843, notes which Gould published in *The Mammals of Australia* (1845–63).

To Europeans it was a creature of enormous interest and beauty, despite its minute size: 150 millimetres from snout to tail-tip, and weight of about 10 grams.

Like the honeyeaters or more particularly the lorikeets, the honey possum is a specialised pollen and nectar eater. To equip it for this life, its teeth have regressed to tiny rudiments, seldom used. Its pointed snout has become greatly elongated and so has its very mobile brush-tipped tongue, which darts out to probe for nectar deep in the base of large blossoms like those of banksias and dryandras.

It is in fact a mammalian counterpart of the feathered honeyeaters, but no one is quite sure who its nearest relatives are. It has some possum-attributes, such as a prehensile tail. Other attributes are more generally marsupial. It may even be distantly related to the kangaroos and wallabies; like them, the female carries dormant blastocysts in her reproductive system, ready to move into the pouch as soon as the pouch-young are weaned, or killed.

To emphasise its uniqueness, the honey possum should perhaps be known simply by the name given it by the Albany Aborigines, 'noolbenger'.

Despite considerable loss of habitat and predation by feral cats and foxes, the noolbenger remains locally common in some sandplain heaths and scrubs of south-western Australia, roughly from Shark Bay to east of Esperance.

Honey possum, *Tarsipes spenserae*

. . . saw them insert their long tongues into the flower

The Tarsipes is generally found in all situations suited to its existence from Swan River to King George's Sound, but from its rarity and the difficulty with which it is procured . . . the natives only brought me four specimens; one of these, a female, I kept alive for several months . . .

It is strictly nocturnal, sleeping during the greater part of the day and becoming exceedingly active at night: when intent upon catching flies it would sit quietly in one corner of its cage, eagerly watching their movements, as, attracted by the sugar, they flew around; and when a fly was fairly within its reach it bounded as quick as lightning and seized it with unerring aim, then retired to the bottom of the cage and devoured it at leisure, sitting tolerably erect and holding the fly between its fore-paws, and always rejecting the head, wings and legs.

The artificial food given it was sopped bread made very sweet with sugar, into which it inserted its long tongue precisely in the way in which the Honey-eaters among birds do theirs into the flower-cups for honey; every morning the sop was completely honey-combed, as it were, from the moisture having been drained from it by the repeated insertion of the tongue; a little moistened sugar on the end of the finger would attract it from one part of the cage to the other; and by this means an opportunity may be readily obtained for observing the beautiful prehensile structure of the tongue, which I have frequently seen protruded for nearly an inch beyond the nose; the edges of the tongue near the tip are slightly serrated . . . Mr Johnson Drummond shot a pair in the act of sucking the honey from the blossoms of the *Melaleuca*; he watched them closely, and distinctly saw them insert their long tongues into the flower precisely after the manner of the birds above-mentioned . . .

Louisa Anne Meredith

Louisa Anne Meredith was by any measure an extraordinarily gifted writer. Born in Birmingham in 1812 and largely educated by her mother, by the time of her marriage in 1839 she had written many reforming political articles and had published three well-received books of verses and prose.

Her reactions to the New South Wales of 1839–40 as an English-woman newly arrived into heat, glare, whirlwinds of dust, flies, cicadas, convict chain-gangs and bushrangers resulted in some of the freshest, most sharply observed passages ever written about the period. In particular, her response to Australian wildlife was spontaneous and sympathetic.

Louisa's husband, Charles Meredith, was to become a landholder and influential, respected politician in Tasmania, where Louisa lived much of her long life. Here she wrote a number of books, which included two novels and some fine natural history.

The excerpts here are from her first Australian book, *Notes and Sketches of New South Wales* (London, 1844), reprinted several times, including a paperback edition in 1973.

Who but a sympathetic naturalist could have made such a picture of our common brushtail possum, *Trichosurus vulpecula*? Clearly Mrs Meredith spent time looking at possums and seems from the accuracy of her picture to have had help from the Aborigines, for whom brushtail possum, singed lightly in its fur to retain the moisture, was a staple food.

So constant was the hunt for possums that despite the inefficiency of stone axes in opening their hollows, the predation must have been an important control on their numbers.

. . . the expression . . . is both like that of the deer and the mouse

A full-grown opossum is larger and heavier than a very large cat, with a pretty innocent-looking face, the expression of which is both like that of the deer and the mouse, the shape of the nose and whiskers strongly resembling the latter.

The eyes are very dark and brilliant, the ears soft and delicate, the legs short and strong, with monkey-like feet and long sharp claws. They sit up, holding their food in the fore-paws, like a monkey . . .

The . . . colour [of the fur is] either black, dark grey, dark brown, or deep golden brown, like very yellow sable, but always beautifully shaded off from the sides towards the under part, which is lighter . . .

Like all other animals of their class, they are marsupial, and have rarely more than one young one at a time, which the doe carries about with her, at first in the pouch, and afterwards on her back.

The blacks procure opossums by climbing trees where their holes are, and have evidently some means of ascertaining whether the animal is turned with its head or tail towards them, before touching it; if the former, they frighten him, or by some means induce him to turn round, when, instantly seizing the tail, they forcibly drag him out.

If the hole extends too far for them to reach their prey, they cut a larger hole with an axe or tomahawk . . . The natives . . . are very agile in climbing trees, making small notches in the bark as they ascend, just large enough to rest the end of the great toe upon, which member seems in them particularly strong . . .

OPPOSITE: Brushtail possum, *Trichosurus vulpecula*, and young

Louisa Anne Meredith

Soon after their arrival in Sydney, Charles Meredith set out to visit his sheep-stations on the Murrumbidgee. Louisa travelled with him as far as Bathurst. There was a severe drought on the land and Louisa's reaction to Bathurst, a month or so after her arrival from England, was not enthusiastic:

. . . the plains of Bathurst . . . being shut in on all sides by lofty ranges of mountains, must endure without any relief of their own oven-like atmosphere . . . when it seems as if a fiery blast from a huge furnace pervaded all space around, rushing into the house through every opening with the force of a hurricane.

My English habit of flinging wide open all doors and windows in warm weather, I here found . . . a most imprudent course to pursue, as the only chance of preserving a moderately endurable existence . . . is, immediately on its approach to shut every door and window, and with closely-drawn blinds to await, as patiently and movelessly as half-suffocated mortals may be expected to do, the abatement of the terrible visitation.

But all was not lost. The Macquarie River still had *some* water in it even in this appalling summer of 1839–40, and here Louisa met the green and golden bell frog, *Litoria aurea*: a tree-frog which, having adapted to climb among vegetation by means of sucking pads on its digits, changed its adaptive mind and now lives in water, as frogs should.

Green and golden frog, *Litorea aurea*

. . . like an embroidery in threads of gold on shaded velvet

In the Macquarie, near Bathurst, I first saw the superb green frogs of Australia. The river, at the period of our visit, was for the most part a dry bed, with small pools in the deeper holes, and in these, among the few slimy water-plants and Confervae, dwelt these gorgeous reptiles.

In form and size they resemble a very large common English frog; but their colour is more beautiful than words can describe. I never saw plant or gem of so bright tints. A vivid yellow-green seems the groundwork of the creature's array, and this is daintily pencilled over with other shades, emerald, olive and blue greens, with a few delicate markings of bright yellow, like an embroidery in threads of gold on shaded velvet.

And the creatures sit looking at you from their moist, floating bowers, with their large eyes expressing the most perfect enjoyment, which, if you doubt whilst they sit still, you cannot refuse to believe in when you see them flop into the delicious cool water, and go slowly stretching their long green legs, as they pass along the waving grove of sedgy, feathery plants in the river's bed, and you lose them under a dense mass of gently waving leaves; and to see this, whilst a burning, broiling sun is scorching up your very life, and the glare of the herbless earth dazzles your agonized eyes into blindness, is almost enough to make one willing to forego all the glories of humanity, and be changed into a frog!

Louisa Anne Meredith

In the Macquarie River at Bathurst there were also platypuses, *Ornithorhynchus anatinus*, which Louisa Meredith described with some astonishment.

The size she gives is rather too short, (average sizes span 40–55 centimetres, or *c.* 16–18 inches) and it is now known that male platypuses *do* have a functional venomous spur in each 'ankle', connected to a venom gland.

If seized in the hand the animal can cling with its hindlegs and drive both spurs inward. Fishermen so wounded by platypuses have suffered swollen limbs and agonies.

Beyond a suggestion that fighting male platypuses may use the spurs against one another when grappling, no clear understanding of the use of the spurs has been reached. An uncomfortable possibility is that they are used, without sufficient force to eject venom, to help hold the female when copulating in the water.

The platypus is still widespread from far south-west Victoria to north-east Queensland; it is common in Tasmania and has been introduced to Kangaroo Island.

On the mainland it is found inland of the Divide as well as in coastal rivers and streams.

Because of its shyness and crepuscular ways it is seldom seen except by fishermen, so that there is an idea that it is very rare, but not so. In some rivers it is common.

Despite complete protection, it does have problems: changes to rivers caused by snagging; increased run-off and erosion from cleared land; seepage of pesticides and herbicides into streams; damaging changes caused by European carp and industrial pollution all mean that the status of the platypus must be carefully watched.

That most enigmatical of all . . . strange animals . . .

That most enigmatical of all the strange animals found in Australia, the Platypus . . . is also a dweller in the Macquarie, but, being extremely shy, is not often found near Bathurst.

So many descriptions have been published of it, that I imagine it is nearly as well understood in England as here. A full-grown specimen is twelve or fourteen inches in length, and much the same shape and proportion as the common mole, with a very thick, soft fur, dark brown on the back, and light coloured beneath; the head and eyes are perfectly *animal*, but in lieu of a mouth or snout, a small flat bill, similar to that of a duck, completes the very odd countenance of this most paradoxical creature.

The short furry legs end in half-webbed feet, the hind ones being armed with sharp spurs, which are perforated, and through which, when the animal is annoyed, it is believed to eject a poisonous fluid as it strikes an enemy; but this fact is still doubted by some naturalists, and, like other anomalous peculiarities, is still the subject of argument amongst the learned, to whose information I regret that it is not in my power to add.

The creature is very rarely seen on shore, and is usually killed by being shot from a high bank; but this is only practicable when it swims very near the surface . . .

Platypus, *Ornithorhynchus anatinus*

John Cotton

Though he came later to Australia than many explorers, John Cotton, born in London in 1801, brought a good knowledge of ornithology and the eye of an artist. Although intended for the law, he had become by degrees a bird-painter and would-be author of fine bird books.

Before moving his family to Australia in middle age, Cotton had already written and illustrated two limited works on the songbirds of Great Britain. How or why he so misjudged his own character or that of the Australian primary industry as to attempt sheep-farming here, we know not.

Settling on his brother's leasehold sheep property 'Doogalook' on the Goulburn River near Mansfield, Victoria, in June 1843, Cotton revealed where his heart's interest lay by recording detailed notes and making more than one hundred excellent water-colours of birds of the district. These became fore-runners for illustrations he hoped to publish in a major book.

Graceful, accurate and true, they reveal him to have been one of the most naturally skilled bird-painters to have worked in this country. But beset by frustrations, Cotton died in 1849 without realising his ambition, although a two-volume work, *Beautiful Birds*, was published in England after his death.

So Cotton's new but informed reactions to the Australian landscape and its birds are doubly interesting. All the birds he mentions in this excerpt are still in the Mansfield district, if one assumes that his 'broadtailed parrots' (a group name for rosellas and their allies) were crimson rosellas, *Platycercus elegans*. ('Rose-hill' was the contemporary name for the beautiful eastern rosella, *P. eximius*.) This short excerpt is from *John Cotton's Birds of the Port Phillip District of New South Wales 1843–1849*: Maie Casey and A. R. McEvey (Sydney, 1974).

All was new and interesting to me . . .

All was new and interesting to me, from the white stems of the lofty Gum, to the deep but lively green of the wattle & native cherry; the honeysuckle trees, acacias and wild flowers.

From the deep sonorous piping song of the magpie to the sharp note or call of the Swift parrakeet, the whistle of the rose-hill & the splendid plumage of the broadtailed parrakeets. The crow of the wattle bird, the amusing chatter of the letherhead or hornbill honey-sucker, the disagreeable scream of the white cockatoo & the curious conglomeration of sounds uttered by the Laughing Jackass . . . by turns attracted my attention.

Noisy friarbird, *Philemon corniculatus*

76

Eastern rosellas, *Platycercus eximius*

Ludwig Leichhardt

Friedrich Wilhelm Ludwig Leichhardt, born in Prussia in 1813, came to Australia in 1841 on a passage paid by an English friend.

The son of a farmer and minor official, Leichhardt had shown academic promise, and for a period was able to attend universities at Gottingen and Berlin. But despite his later assumption of the title of 'Dr', seems never to have taken a degree.

In Australia, after some desultory 'scientific' investigations and travelling, Leichhardt learned there was interest among Queensland squatters about pastoral prospects in the unknown north.

At Sydney he somehow achieved support for an exploration which would take him some 5000 kilometres through sub-coastal northern Australia to the settlement of Port Essington in Arnhem Land. Leichhardt assembled a mixed party at Westbrook Station, near Toowoomba on the Darling Downs. From there, in September 1844, he set off.

At the last moment Leichhardt accepted the application of John Gould's field-collector John Gilbert, who became Leichhardt's second in command.

Much has been written about the hardships, bungling and beginner's luck of Leichhardt's party. The thing that finally mattered was that in December 1845, in a little over fourteen months, they got through to Port Essington, discovering much useful land and nearly a dozen major rivers *en route.*

The feat, certainly Leichhardt's major success in life, made him famous and temporarily prosperous. But the glory was brief: on his next expedition-but-one, his party disappeared without trace while trying to cross northern inland Australia.

These excerpts are from Leichhardt's *Journal of an Overland Expedition in Australia from Moreton Bay to Port Essington* (London, 1847).

On 17 March 1845, travelling north along the valley of north Queensland's Suttor River, Leichhardt's party came unexpectedly on Lake Suttor.

In open country, wrote Leichhardt, the practised eye could pick up leading signs to water: a cluster of trees of greener foliage; cockatoos and pigeons gathering, especially towards sunset; the calls of magpie-larks and flocks of finches, would all claim attention. But when travelling through thick brigalow scrub it was mostly a matter of chance. So they were delighted with this find.

Such remote water-lily-decked havens are often thronged with waterbirds in Queensland even today. Crowds of plumed whistling ducks, *Dendrocygna eytoni,* camp on their banks, moving out into grasslands to feed at dusk. Gorgeous green pygmy geese, *Nettapus pulchellus,* the most beautiful of small ducks, go among the blue haze of water-lily blossoms on the dark water. Cormorants and darters, *Anhinga melanogaster,* dry their wings on stumps or feed young in nests over water.

Plumed whistling ducks, *Dendrocygna eytoni*

Swarms of ducks covered the margin of the lake . . .

Charley cried out, "Look there, Sir! what big water!" and a long broad
sheet of water stretched in sweeps through a dense Bauhinia and Bricklow
scrub, which covered the steep banks.

It is a singular character of this remarkable country, that extremes so
often meet; the most miserable scrub, with the open plain and fine forest
land; and the most paralysing dryness, with the finest supply of water.

Swarms of ducks covered the margin of the lake; pelicans, beyond the
reach of shot, floated on its bosom; land-turtles plunged into its waters;
and shags started from dead trees lying half immersed, as we trod the well-
beaten path of the natives along its banks.

The inhabitants of this part of the country, doubtless, visit this spot
frequently, judging from the numerous heaps of muscle-shells. This fine
piece of water, probably in the main channel of the Suttor, is three miles
long, and is surrounded with one mass of scrub, which opens a little at its
north-western extremity . . .

John Gilbert

There is pathos in these extracts from John Gilbert's diary, even so long after the event. (Diary of the Leichhardt Expedition to Port Essington, 1844–1845; Mitchell Library.)

Hours after writing that last entry on 28 June 1845, Gilbert was killed by a spear as Aborigines ambushed the camp of the Leichhardt expedition, near the Gilbert River on the western base of Cape York Peninsula.

Leichhardt's diary says nothing about the cause of the attack—the only serious one the party sustained in the fourteen month trek from the Darling Downs to Port Essington, on the north coast of Arnhem Land. But later the story came out. The attack was apparently made in retribution for assault on an Aboriginal woman or women by Charley and Brown, two blacks with the Leichhardt party.

And so, quite needlessly, died John Gilbert, probably our greatest ornithological explorer and collector—a man sympathetic to the Aborigines and their way of life. He had gamely gone with Leichhardt simply for the chance to fill the gaps in Gould's coverage of the birds of Australia.

After the tragedy, Leichhardt carried Gilbert's diary through to Port Essington and sent it to Gould.

Since Gilbert died the numbers of black kites, *Milvus migrans*, in northern Australia have probably been swollen by pastoral changes and access to cattle-camp offal.

Whistling ducks, both plumed, *Dendrocygna eytoni*, and diving, *D. arcuata*, are still widespread, as are the other waterfowl. But their numbers have probably greatly declined overall because of grazing and the removal of swamp-vegetation by cattle.

In general, cattle, and in the Top End, buffaloes, have wrought great changes to the country Gilbert described.

... the Kites as numerous as ever

Fri. 27 June, 1845. [The black kite] ... is on the increase as well in numbers as in boldness. In the afternoon while sitting at the entrance of my tent skinning birds I had a tin case with specimens between my legs, the lid of which I had opened to air the specimens enclosed, among which was the only specimen of my last new Honey-sucker.

This was lying on the top and had deceived the bird so much that he darted down, and to my surprise and vexation fairly carried off my specimen, and flying into a neighbouring tree instantly plucked it to pieces, whether he swallowed any I could not tell, but at all events I should imagine that the Arsenic will not at all agree with its stomach although they display a little nicety in what they pick up ...

Saturday 28 ... We had again rather a change of country on crossing the creek, we entered a finer forest than we have met with for some time past, the timber consisting principally of Stringy Bark, Box and Bloodwood, and very fine grass, from this we entered a flat wet country again, at about four miles we crossed a considerable creek, ... from this the remaining part of the stage was through a beautiful open country thickly studded with lotus ponds at one of which we camped. Native fires in every direction and very near us, but none of the natives seen ... [*Dendrocygna* ...] again abundant. Brown killed 6 at a shot. The Wood Duck, Teal and Black Duck still abound, and the Kites as numerous as ever in fact we have marked several of them and seen them again and again at succeeding camps, so that there is no doubt that they regularly follow us from place to place as do the crows, which we a long time ago remarked ...

Black kite, *Milvus migrans*

Ludwig Leichhardt

By early October 1845, Leichhardt's party had nearly got past the difficult country along the south-west corner of the Gulf of Carpentaria. The weather was very hot. Their clothes were in tatters and Leichhardt's straw hat had caught fire, obliging him to wear a strange head-dress made of a canvas bag.

Except for strips of dried beef and greenhide, they were now living off the land: on emus (smaller and leaner than in New South Wales); wallabies, flying foxes, fish and on fruits and seeds they had learned to crush, wash or heat to avoid explosive results in the bowels. But none of them was getting fat.

Crossing the remote Limmen Bight River in the south-west corner of the Gulf, Leichhardt struck north-west and soon found another river, fresh this time, with verdant green surroundings. He named this the Wickham. (It is now the Cox.)

In these new hopeful surroundings, the party found plenteous wildlife, particularly red or sandy wallabies, *Macropus agilis*, and a much more notable animal, called by Leichhardt the 'red forester'—now known as the antilopine wallaroo, *Macropus antilopinus*.

This was a prize indeed. True to its name a rangy, powerful and really striking creature, it approaches the red and grey kangaroos in size and is often spoken of as one of the five 'great' kangaroos.

Antilopine wallaroos live mostly in the grassy, open tropical woodlands and foothills of Cape York Peninsula, the top end of the Northern Territory and the Kimberleys. Their parties move about feeding in the mornings and evenings, when their very distinctive reddish or greyish buff coats, paling to almost white below and on the limbs, make them most attractive to see in these savannah-like surroundings.

Sandy wallabies, *Macropus agilis*

. . . we started a flock of red foresters

We crossed the river and travelled about ten miles north-west, over a succession of stony ridges, separated by fine open tea-tree and box flats . . .

At the end of the stage, the uniform colour of the country was interrupted by the green line of a river-bed, so pleasing and so refreshing to the eye, with the rich verdure of its drooping tea-trees and myrtles, interspersed with the silver leaves of *Acacia neurocarpa*, and *Grevillea chrysodendron*.

The river was formed by two broad sandy beds, separated by a high bergue and was full 700 yards from bank to bank. It contained large detached water-pools fringed with Pandanus, which were very probably connected by a stream filtering through the sands, I called it the "Wickham," in honour of Captain Wickham, R. N. of Moreton Bay . . .

The red wallabi . . . was very numerous along the gullies of the river: and we started a flock of red foresters . . . out of a patch of scrub on the brow of a stony hill. Charley and Brown, accompanied by Spring, pursued them, and killed a fine young male.

I had promised my companions that, whenever a kangaroo was caught again, it should be roasted whole, whatever its size might be. We had consequently a roasted Red Forester for supper, and we never rolled ourselves up in our blankets more satisfied with a repast . . .

Antilopine wallaroo, *Macropus antilopinus*

Ludwig Leichhardt

Four days after their feast of 'red forester' the Leichhardt party was pushing through rough, scrubby country round the corner of the Gulf, where cypress-pine thickets alternated with stringybarks. Sunset of an exhaustingly hot day (17 October 1845) found them preparing to camp on a dry, open plain.

Someone knocked over their water-pot and they would have faced a thirsty night had not Charley, an Aboriginal member of the party, discovered waterholes in a small creek nearby. At moonrise they shifted camp and the following day rested beside these pools, giving Leichhardt time for rest in an environment that could only make a modern birdwatcher drool.

Where undamaged by cattle, such waterholes still attract swarms of birds like diamond doves, *Streptopelia cuneata*, and kin, and also honeyeaters. The 'little finches' probably included the long-tailed finch, *Peophila acuticauda*, the crimson finch, *Neochmia phaeton*, and possibly the gorgeous gouldian finch, *Erythrura gouldiae*, now much rarer in northern Australia possibly because of loss of habitat, illegal trapping and other factors not certainly understood.

Diamond dove, *Streptopelia cuneata*

Gouldian finch, *Erythrura gouldiae*

It was highly amusing to watch the swarms of little finches . . .

I stopped at the water-holes, to allow our cattle to recover. It was a lovely place. The country around us was very open, and agreeably diversified by small clusters of the raspberry-jam tree. Salicornia and Binoe's Trichinium indicated the neighbourhood of salt water; but the grass was good and mostly young . . .

The water-hole on which we were encamped was about four feet deep, and contained a great number of guard-fish, which, in the morning, kept incessantly springing from the water. A small broad fish with sharp belly, and a long ray behind the dorsal fin, was also caught.

It was highly amusing to watch the swarms of little finches, of doves, and [honeyeaters], which came during the heat of the day to drink from our water-hole.

[Magpie Larks], Crows, Kites, Bronze-winged and [Flock] pigeons . . . The Rose cockatoo . . . the Betshiregah . . . and [the Varied Lorikeet] . . . were also visitors to the water-hole, or were seen on the plains. The day was oppressively hot; and neither the drooping tea-trees, nor our blankets, of which we had made a shade, afforded us much relief . . .

Ludwig Leichhardt

The immense, sandstone escarpments of Arnhem Land, with patches of tropical rainforest clinging to their fissures and gorges, are a very distinctive environment.

It was this wild region that Leichhardt's party now had to cross. It was immensely trying in the gathering heat of tropical summer, but was enlivened here and there by sight of distinctive Arnhem Land wildlife.

They encountered the sandstone shrike-thrush, *Colluricincla woodwardi*, a handsome creature well-adapted to life on the rock-faces, which throw back its splendid voice in repeated, ringing echoes.

Short-sighted Leichhardt may not have actually *seen* the bird, but his description is accurate: 'It raised its voice, a slow full whistle, by five or six successive half-notes; which was very pleasing, and frequently the only relief while passing through this most perplexing country.'

Just before their release from this beautiful but terrifying maze, they met the chestnut-quilled rock-pigeon, *Petrophassa rufipennis*, a bird nearly confined to the Arnhem Land escarpment, whose very appearance and function have been shaped to match the dark, granular rocks of its home.

Plump and low-slung, it moulds to the rock. Its plumage is mostly the same reddish charcoal colour as the substrate, but about its head and neck the feathers each have a tiny touch of white, precisely resembling the rock's granular appearance.

Soon after this discovery, Leichhardt's luck re-asserted itself. Suddenly, on 17 November 1845, 'the extensive view of a magnificent valley opened before us. We stood with our whole train on the brink of a deep precipice, of perhaps 1800 feet descent . . . A large river . . . meandered through the valley . . .'

It was the valley of the South Alligator River, their pathway out of the wilderness.

A new species of rock pigeon . . .

We accomplished about ten miles in a direct line, but on a long and fatiguing circuitous course. . . . At the distance of four miles I came to a rocky creek going to the westward, which I followed.

From one of the hills which bounded its narrow valley, I had a most disheartening, sickening view over a tremendously rocky country.

A high land, composed of horizontal strata of sandstone, seemed to be literally hashed, leaving the remaining blocks in fantastic figures of every shape; and a green vegetation, crowding deceitfully within their fissures and gullies, and covering half of the difficulties which awaited us on our attempt to travel over it.

The creek, in and along the bed of which we wound slowly down, was frequently covered with large loose boulders, between which our horses and cattle often slipped.

A precipice, and perpendicular rocks on both sides, compelled us to leave it; and following one of its tributary creeks to its head, to the northward, we came to another, which led us down to a river running to the west by south.

With the greatest difficulty we went down its steep slopes, and established our camp at a large water-hole in its bed . . .

A new species of rock pigeon . . . with a dark brown body, primaries light brown without any white, and with the tail feathers rather worn, lived in pairs and small flocks . . . and flew out of the shade of overhanging rocks, or from the moist wells which the natives had dug in the bed of the creek, around which they clustered like flies round a drop of syrup . . .

The river was densely covered with scrub, and almost perpendicular cliffs bounded its valley on both sides. Myriads of flying-foxes were here suspended in thick clusters on the highest trees in the most shady and rather moist parts of the valley. They started as we passed, and the flapping of their large membranous wings produced a sound like that of a hail-storm . . . During the night we heard the first grumbling of thunder since many months . . .

Chestnut-quilled rock pigeon, *Petrophassa rufipennis*

Ludwig Leichhardt

Pied geese, *Anseranas semipalmata*

After escaping the Arnhem Land escarpment, Leichhardt's small, lightly-equipped party reached Port Essington on Cobourg Peninsula in December 1845, less than fifteen months after leaving the Darling Downs, 5000 kilometres behind. Leichhardt's luck held to the end: even to the wet season being delayed long enough to let them safely cross the great northern coastal plain.

The same late start to the wet season meant that the almost unbelievable concentrations of water-birds on the remaining coastal wetlands had not dispersed. Leichhardt's vision of armies and phalanxes of pied geese, *Anseranas semipalmata*, whistling ducks, 'white cranes' (egrets), brolgas and other water-birds as well as red-tailed black cockatoos, *Calyptorhynchus magnificus*, spell out the richness that was there.

Soon the glory of the tropical lagoons was devastated by large herds of introduced water-buffaloes which strip vegetation from the wetlands, leaving thousands of hectares of bare, pugged mud.

. . . the noise of clouds of water-fowl . . . betrayed to us the presence of water

About six miles from our last camp, an immense plain opened before us, at the west side of which we recognized the green line of the river.

We crossed the plain to find water, but the approaches of the river were formed by tea-tree hollows, and by thick vine brush, at the outside of which noble bouquets of Bamboo and stately Corypha palms attracted our attention.

In skirting the brush, we came to a salt-water creek (the first seen by us on the north-west coast), when we immediately returned to the ridges, where we met with a well-beaten foot-path of the natives, which led us along brush, teeming with wallabies and through undulating scrubby forest ground to another large plain.

Here the noise of clouds of water-fowl, probably rising at the approach of some natives, betrayed to us the presence of water. We encamped at the outskirts of the forest, at a great distance from the large but shallow pools, which had been formed by the later thunder-showers . . .

. . . Since the 23rd of November, not a night had passed without long files and phalanxes of geese taking their flight up and down the river, and they often passed so low, that the heavy flapping of their wings was distinctly heard. Whistling ducks, in close flocks, flew generally much higher, and with great rapidity.

No part of the country we had passed, was so well provided with game as this; and of which we could have easily obtained an abundance, had not our shot been all expended.

The cackling of geese, the quacking of ducks, the sonorous note of the native companion, and the noises of black and white cockatoos, and a great variety of other birds, gave to the country, both night and day, an extraordinary appearance of animation . . .

Richard Howitt

Perhaps Richard Howitt (1799–1869) contributed less to the Australian colonies in his four years here (1840–44) than his medico brother Godfrey and his explorer-scientist nephew, Alfred Howitt.

But his descriptions of countryside and wildlife, people and manners, particularly those in his book *Impressions of Australia Felix* (London, 1845), throw a clear and loving light on a land being rudely dragged from a settled way of ages.

The first entry describes the commencement of clearing a 40-hectare property the Howitts bought in June 1840 on the north side of the Yarra near present Alphington, a few kilometres from the centre of Melbourne. Now, the area is an industrial and residential suburb. Then, it was a handsome landscape of immense river redgums, *Eucalyptus camaldulensis*.

'The situation is delicious', he wrote, the soil tolerably rich; the slopes most graceful. The windings of the Yarra in full prospect, both near and far-off, are beautiful'.

But then he and his brother started clearing part of the block for crops and pasture. And here Richard's poetic soul was devastated at the destruction they had to cause. His heart was not in it, and he soon went wandering again.

He left Australia to be tamed and—in the sense of wildlife and native plants—depauperised by settlers less sensitive than himself.

Howitt was not in error in describing young 'wild cats' (eastern quolls, *Dasyurus viverrinus*), as 'black creatures spotted with white'. There were two colour-forms: one fawn-brown, the other, blackish; both had white spots on the body.

Howitt's 'flying squirrels' were probably sugar gliders, *Petaurus breviceps*, still present along the Yarra in places.

. . . trees . . . of unconscionable girth

Day after day it was no slight army of trees against which we had to do battle; . . . for the trees in those days were giants . . . There was a world of work, digging to lay bare the roots, felling, and then cutting the boles and boughs up with the saw and axe. Such of the boles as were good for anything we cut into proper lengths for posts; splitting and mortising them for that purpose. Rails also we had to get when there were any boughs straight enough.

Some of the trees were of unconscionable girth, six or eight yards in circumference. Immense was the space of ground that had to be dug away to lay bare the roots. And then, what roots! they were too large to be cut through with the axe; we were compelled to saw them in two with the cross-cut saw.

One of these monsters of the wild was fifteen days burning; burning night and day, and was a regular ox-roasting fire all the time.

We entirely routed the quiet of that old primeval forest solitude, rousing the echo of ages on the other side of the river, that shouted back to us the stroke of the axe, and the groan and crash of falling gum-trees . . .

Then what curious and novel creatures,—bandicoots, flying squirrels, opossums, bats, snakes, guanas, and lizards—we disturbed, bringing down with dust and thunder their old domiciles about their ears. Sometimes, also, we found nests of young birds and of young wild cats; pretty black creatures, spotted with white.

The wild denizens looked at us wildly, thinking, probably, that we were rough reformers, desperate radicals, and had no respect for immemorial and vested rights. It was unnatural work, and cruel . . . The horrid gaps and blank openings in the grand old woods seemed . . . to reproach us.

It was reckless waste, in a coal-less country, to commit so much fuel to the flames. Timber, too, hard in its grain as iron almost, yet ruddy, and more beautiful than mahogany. No matter, we could not eat wood; we must do violence to our sense of the beautiful, and to Nature's sanctities; we must have corn-land, and we, with immense labour, cleared seventeen acres. . . .

OPPOSITE: Sugar glider, *Petaurus breviceps*

88

Richard Howitt

In the summer of 1843–44, when Melbourne was but eight years old, Richard Howitt made a walking trip from Melbourne to Westernport, then to Arthur's Seat and Cape Schanck, before returning.

He left Melbourne on 29 December 1843 and walked, by way of the charming new seaside resorts of St Kilda and Brighton, towards Major James Frazer's run near Mordialloc.

In early January, on his way back through Willoughby's Station east of present Pearcedale, Howitt fell in with a bird. Today's residents of that district will have no trouble recognising the strident creature still common there, and still aggressive. The noisy miner, *Manorina melanocephala*, is a honeyeater and not to be confused with the introduced Asian common mynah, *Acridotheres tristis*, one of the starlings.

Living in aggressive, co-operative colonies that drive out other birds, noisy miners instantly want to know the business of any human entering the group-territory and frequently fly shrilling loudly toward a stranger or hang berating him from a branch. European settlement has probably assisted noisy miners, which prefer open woodlands, by clearing and opening gaps in the former forests and by feeding them at bird-tables.

. . . the very sentinel of the woods

My next resting place, after six or eight miles' walk, was Willoughby's Cattle Station; and the whole of the way the country is of one character—covered over with deep heather—thousands, or we might almost say millions, of acres, of worthless forest—the scrub and trees stunted, stringy bark . . .

Hence to Allen's Station was two miles. Betwixt these stations I met with a new acquaintance—a grey bird, the size of a thrush . . . with pale yellow bill and legs, flew to meet me, and settled familiarly just before my face, and looked inquiringly as to the reason of my being there. The shrewd look it gave me, made me think of the enchanted birds in "The Arabian Nights".

I was surprised at the creature's freedom and boldness, but immediately saw the reason of it—a nest, a little further on, depended in a bough over the road. This I drew down with my stick, and saw in the nest two salmon-coloured spotted eggs.

But, the disturbance there was immediately! The whole wood was in one clamour of resentment. My new acquaintance of the nest commenced the outcry, and I shall never forget how first one, and then another, took it up, and continued it, 'till all the forest rang with the sound. "Shrill! shrill! shrill! shrill!" was the sharp, quick, iteration everywhere, and all at once . . .

I found that this creature was very appropriately named the soldier-bird. It is the very sentinel of the woods, sending far on before you intelligence of your coming . . .

Noisy miner, *Manorina melanocephala*

Richard Howitt

After leaving Willoughby's Station on Mornington Peninsula, Richard Howitt pressed on east to his destination, Manton's Station near present Tooradin.

On his return he headed toward Captain Baxter's Station below Mt Eliza. But by taking a 'deceitful cattle track' he became temporarily lost in the bush and had to make a makeshift hut for the night.

Howitt's description of his night echoes the experience of many early bush travellers, benighted between stations. The dingo, *Canis familiaris dingo*, is now long-gone from Mornington Peninsula. The 'kangaroo rat' was probably the long-nosed potoroo, *Potorous tridactylus*, nearly wiped out locally by clearing, foxes and cats.

The flying squirrel was the sugar glider, *Petaurus breviceps*, still common where there are suitable trees. The possums, by their calls, were doubtless brushtails, *Trichosurus vulpecula*.

The snake—who can tell? But copperheads, *Austrelaps superbus*, and tiger snakes, *Notechis scutatus*, are both still common in the district. The latter, particularly, are sometimes active on warm summer nights.

Long-nosed potoroo, *Potorous tridactylus*

My bed I . . . made of branches . . .

Glad was I at length to be once more on the cart-track . . .

On the way I recollected seeing where the bark-peelers had built, and thrown down again, a temporary miam, as the natives would call it. There I determined, . . . to pass the night.

On I went, and without any concern saw the sun gradually go down, the dark to gather round me, miles from any habitation, and the moon and stars grow bolder and brighter . . .

I set up the cast-down poles—one end in a forked tree—the other resting on the ground. These I thatched over with the ready-cut branches of the wild cherry-tree: and had soon a very snug house. My bed I next made of branches of a shrub, very myrtle-like, and of heath.

Then, with the fire blazing brightly at my feet, with my carpet-bag for a pillow, wrapped warmly in my blanket, the laughing-jackass merrily bade me good-night, and I slept soundly at intervals—waked sometimes by the melancholy howl of a wild dog, or a rustle amongst the leaves of my house of perhaps a snake or of a kangaroo rat—to hear a little way from me in the trees—plop-plop—the noise of the flying squirrel going from bough to bough: and the sharp guttural noises of opossums.

Then what a hush amidst a gentle breeziness would come over the wilderness? Soft as feathers was the dry and balmy atmosphere—the moon hanging how amazingly near me, like a large pearl, and the stars, as near, like intense fiery rubies . . .

John Macgillivray

The debt Australian zoology owes the British navy is seldom realised. At intervals for more than a century, from 1770 on, the Admiralty had ships exploring and surveying the Australian coast and its offshore waters.

Those ships frequently carried official botanists or zoologists or there were competent naturalists among their officers, who gathered specimens for museums and private collections.

Sir Joseph Banks and later John Gould had many specimens from this source. In this way His (or Her) Majesty's ships, *Endeavour, Investigator, Norfolk, Lady Nelson* and later the *Beagle*, the *Fly* and the *Rattlesnake* all made significant contributions to zoological discovery in Australia.

John Macgillivray (1821–67), son of an eminent Scottish zoologist, became a medical student at the University of Edinburgh. In a manner not unknown today, he then switched to zoology and is next heard of in 1842 coming to Australia on the surveying corvette HMS *Fly* as naturalist-collector to the Earl of Derby, the Lords Commissioners of the Admiralty having given their approval.

The *Fly* and her successor, HMS *Rattlesnake*, made a series of long surveys on the Australian and New Guinea coasts in the years 1842–50, giving Macgillivray a magnificent opportunity to see much wildlife of the coastal regions.

The *Fly* spent June 1844 near Raine Island, where a beacon was erected to help guide shipping through the reefs to Torres Strait. (Raine Island is on the Great Barrier Reef about 150 kilometres off the north-east Queensland coast, E.N.E. of Cape Grenville.) Macgillivray has left a record of stupendous seabird life that was there.

Remote and isolated, later in the 19th century the island attracted guano-gatherers who dug out the age-old deposits formed by seabirds in earlier geological times. As well as this exploitation and the staggering toll taken by the *Fly*'s people, unknown numbers of seabirds and their eggs have since been taken from Raine Island by foreign fishermen and by Torres Strait islanders. Despite the exploitation, the seabirds are still a spectacle, if no longer the stupendous one Macgillivray saw.

Birds recorded by Macgillivray which still breed on Raine Island include: the masked, brown, and red-footed boobies, — *Sula dactylatra; S. leucogaster* and *S. sula*; the lesser frigatebird, *Fregata ariel*; small numbers of the beautiful red-tailed tropicbird, *Phaethon rubricauda*; (the only breeding station close to the Australian east coast); the common noddy, *Anous stolidus*; and the Caspian and sooty terns, respectively *Sterna caspia* and *S. fuscata*. This excerpt is from Macgillivray's *Narrative of the Voyage of HMS Rattlesnake*, 2 vols (London, 1852).

... one of those vast breeding-places of birds

We found not less than eighteen species of birds, several of which were new to science, inhabiting this mere speck of land ...

Upon nearing the shore on my first visit in May, 1844, an immense cloud of sea-fowl was observed hovering over the place, and their cries were distinctly heard at a distance of a mile.

Crossing the reef, we landed on a steep sandy beach, and a few yards further brought us upon one of those vast breeding-places of birds, of which none but an eye-witness can form an adequate idea.

The ground was so thickly strewed with eggs, that we could not walk about without occasionally crushing them underfoot; myriads of terns, noddies and boobies darkened the air around; the mingling of loud, harsh, discordant cries was absolutely deafening, and caused even a painful sensation, which, with the stench from numbers of putrifying carcasses and other sources, was almost insufferable.

The birds appeared so little accustomed to the sight of man, that many, busily engaged in incubation, allowed of very close approach. Some frigate-birds sitting upon their nests awaited our coming up with perfect composure, and stoutly defended their eggs with open beak, nor would they take to wing until pushed off the nest with a stick.

A large flock of gannets and boobies covered a bare spot in the centre of the island, chequered black and white with their dense masses.

The eggs and newly-fledged young of the tern and noddy were turned to good account by the party established upon the island, and with an occasional turtle, now and then some fish, and abundance of fresh vegetables, they fared considerably better than on board ship. I amused myself one day with making a calculation of the consumption of young birds and eggs during the month of June, and found it to amount, at the lowest reasonable estimate, to 3000 of the former, and 1410 dozen of the latter ...

Brown boobies, *Sula leucogaster*

John Macgillivray

The *Rattlesnake* was sent out (with the *Bramble*) in 1847 to continue the *Fly*'s surveying work; Macgillivray again was permitted to travel on her, under the rather tragic captaincy of Owen Stanley, who committed suicide on the ship in Sydney in March 1850.

This excerpt, and the following, are from Macgillivray's *Narrative of the Voyage of HMS Rattlesnake* (London, 1852, 2 vols).

Up and down the east coast of Australia, and elsewhere, stories have long been told of an almost mystical partnership between Aborigines and dolphins of one species or another, when fish were running.

It seems certain that, in parts, a mutually profitable relationship did grow up between these highly unlikely partners, for good reasons: both would benefit from shoaling fish being hunted cooperatively.

No doubt it all seems far-fetched, but if killer whales could drive baleen whales into the reach of shore-whalers at Twofold Bay, surely dolphins and Aborigines, who had much longer to become associated, might have done something similar at places like Moreton Island, where Macgillivray observed it in the late 1840s.

From another reference Macgillivray makes to the subject, the Moreton Bay Aborigines highly and superstitiously valued their friends the dolphins. When Macgillivray had earlier sought help from the Aborigines to secure a specimen of this dolphin for his collection he was quite unsuccessful, 'as the natives believed the most direful consequences would ensue from the destruction of one'.

... on the best possible terms with the porpoises

We took up our former anchorage under Moreton Island and remained there for nine days, ...

One night while returning from an excursion, I saw some fires behind the beach near Cumboyooro Point, and on walking up was glad to find an encampment of about thirty natives, collected there for the purpose of fishing, this being the spawning season of the mullet, which now frequent the coast in prodigious shoals.

Finding among the party an old friend of mine, usually known by the name of Funny-eye, I obtained with some difficulty permission to sleep at his fire, and he gave me a roasted mullet for supper.

The party at our bivouac, consisted of my host, his wife and two children, an old man and two wretched dogs. We lay down with our feet towards a large fire of drift wood, partially sheltered from the wind by a semicircular line of branches, stuck in the sand behind us; still, while one part of the body nearly roasted, the rest shivered with cold.

The woman appeared to be busy all night long in scaling and roasting fish, of which, before morning, she had a large pile ready cooked; neither did the men sleep much—for they awoke every hour or so, gorged themselves still further with mullet, took a copious draught of water, and wound up by lighting their pipes before lying down again.

At daylight every one was up and stirring, and soon afterwards the men and boys went down to the beach to fish. The rollers coming in from seaward broke about one hundred yards from the shore, and in the advancing wave one might see thousands of large mullet keeping together in a shoal with numbers of porpoises playing about, making frequent rushes among the dense masses and scattering them in every direction.

Such of the men as were furnished with the scoop-net waded out in line, and, waiting until the porpoises had driven the mullet close in shore, rushed among the shoal, and, closing round in a circle with the nets nearly touching, secured a number of fine fish, averaging two and a half pounds weight. This was repeated at intervals until enough had been procured.

Meanwhile others, chiefly boys, were at work with their spears, darting them in every direction among the fish, and on the best possible terms with the porpoises, which were dashing about among their legs, as if fully aware that they would not be molested ...

OPPOSITE:
Bottle-nosed dolphins, *Tursiops truncatus*

John Macgillivray

On 1 October 1849 the *Rattlesnake* arrived at Evans Bay, Cape York, for a final visit before returning to Sydney, where the tour of duty ended.

Cape York Peninsula and New Guinea share many similarities in climate, vegetation and wildlife. In effect they form a faunal unit divided about 8000 years ago after the last Pleistocene ice-sheets melted and raised ocean levels round the world.

Among other large-scale changes this caused in the world map, the inundation created Torres Strait and separated New Guinea from Australia. So today certain birds, mammals, reptiles, amphibians and insects are found both in New Guinea and on Cape York but nowhere else in Australia.

Macgillivray found his new and interesting prizes in birds of this group. They include some that still migrate between the two countries and some that are sedentary in both.

Neither the palm cockatoo, *Probosciger aterrimus*, nor the fawn-breasted bowerbird, *Chlamydera cerviniventris*, is known to cross Torres Strait and bridge the two populations.

But the gorgeous white-tailed kingfisher,* *Tanysiptera sylvia*, has a different strategy which it was possibly following long before the watery separation of the lands. A warmth-loving creature, it makes the long sea-crossing to New Guinea in March and April each year and winters there, returning to lowland rainforests, south to near Townsville, in October and November, to breed in tunnels drilled in small, round termite mounds.

So these finds by Macgillivray on Cape York, thrilling in themselves because each of the birds is spectacular in its way, had greater significance. They became part of a growing evidence that our continent has not always been as it is today.

* See title page.

Among . . . the ornithological collections . . . were . . . new species of birds, and . . . others previously known only as inhabitants of New Guinea . . .

Among many additions to the ornithological collections of the voyage were eight or nine new species of birds, and about seven others previously known only as inhabitants of New Guinea and the neighbouring islands.

The first of these which came under my notice was an enormous black parrot . . . with crimson cheeks; at Cape York it feeds upon the cabbage of various palms, stripping down the sheath at the base of the leaves with its powerful, acutely-hooked upper mandible.

The next in order of occurrence was a third species of the genus *Tanysiptera* . . ., a gorgeous kingfisher with two long, white, central tail-feathers, inhabiting the brushes, where the glancing of its bright colours as it darts past in rapid flight arrests the attention for a moment ere it is lost among the dense foliage . . .

Two days before we left Cape York I was told that some bower-birds had been seen in a thicket . . . half a mile from the beach, and after a long search I found a recently constructed bower, four feet long and eighteen inches high, with some fresh berries lying upon it.

The bower was situated near the border of the thicket, the bushes composing which were seldom more than ten feet high, growing in smooth sandy soil without grass.

Next morning I was landed before daylight, and proceeded to the place in company with Paida, taking with us a large board on which to carry off the bower as a specimen.

I had great difficulty in inducing my friend to accompany me, as he was afraid of a war party of Gomokudins, which tribe had lately given notice that they were coming to fight the Evans Bay people. However I promised to protect him, and loaded one barrel with ball, which gave him increased confidence, still he insisted upon carrying a large bundle of spears and a throwing stick . . .

While watching in the scrub I caught several glimpses of the tervinya (the native name) as it darted through the bushes in the neighbourhood of the bower, announcing its presence by an occasional loud churr-r-r-r, and imitating the notes of various other birds, especially the leather-head.

Fawn-breasted bowerbird, *Chlamydera cerviniventris*

Palm cockatoos, *Probosciger atterimus*

I never before met with a more wary bird, and for a long time it enticed me to follow it to a short distance, then flying off and alighting on the bower, it would deposit a berry or two, run through, and be off again . . .

At length, just as my patience was becoming exhausted, I saw the bird enter the bower and disappear, when I fired at random through the twigs, fortunately with effect . . .

My bower-bird proved to be a new species . . . *Chlamydera cerviniventris* . . . the bower is [now] exhibited in the British Museum . . .

H. W. Wheelwright

Of all the explorers great and small, settlers, pastoralists; journal-keeping soldiers and sailors, naturalists, scientists, collectors and others who reported natural Australia in the past two centuries, none saw it quite as Horace Wheelwright did.

The son of an Anglican rector, well-read as sons of country clergy often were, he was born in Northamptonshire in 1815 and educated at Reading Grammar School. Intended for the law, he practised for a few years as a solicitor. But the natural world and field sports were his only interest and he soon went wandering in the wilds in Norway and Sweden.

In 1852 Wheelwright came to Victoria to live for four years as a professional wildfowler, based—of all places—near Mordialloc Creek on Melbourne's bayside, at the height of the Ballarat gold rush. Styling himself 'An Old Bushman', he later recorded his Australian experiences in a celebrated small book *Bush Wanderings of a Naturalist* (London, 1861), probably the most complete account of the wildlife of an Australian district published last century.

'Six years' rambling over the forests and fells of Northern Europe had . . . unfitted me for any settled life' Wheelwright wrote in his introduction: 'I had no luck in the diggings. The town was out of the question, and to keep the wolf from the door there were but two alternatives, to seek work on a station, or face the bush on my own account. I chose the latter, and never regretted that choice.'

Mordialloc is now a populous Melbourne outer suburb, far from wild. But in the 1850s the bayside hinterland was mostly covered by the vast Carrum Carrum swamps. Here Wheelwright and his nameless English mate camped and shot.

In those days before sophisticated food-distribution and refrigeration, wildfowling for the market was an accepted source of supply of game. The drawbacks in Australia to this trade were distance, heat and blowflies. Wheelwright wrote:

When I . . . commenced shooting, Melbourne was our only mart, and as we had then no horse, and no means of getting the game up than carrying it ourselves, a journey [of some 25 kilometres] into town on foot at night, with a heavy swag of game (for the small game in the summer must be sold early in the day after it is killed) after a hard day's shooting, was no joke . . . There was no big game in Australia . . . But for small game, I don't think this country can be surpassed; and ducks, pigeon, quail and snipe can be killed in almost any quantities, at the proper seasons . . .

In support of this extravagant claim Wheelwright quoted from his game book of 22 December 1854:

. . . all was game on that day. At night we brought home to my tent— 16½ couple quail, 3½ couple scrub quail, 1 rail, 3 couple pigeons, 11 couple snipe, 3 nankeen cranes, 1 red lowry, 5 black-ducks, 3 shovellers, 3 coots, 2 black cockatoos, 2 moorhens, 7 shell parroquets.

Clearly they shot anything that got its feet off the ground and much that did not. The scrub-quail were probably painted buttonquail, *Turnix varia*, and the shell paroquets were of course budgerigars, *Melopsittacus undulatus*.

Even that level of slaughter could have been maintained had habitats been preserved. Step by step though, over the next seventy years, the Carrum Carrum swamps, and others, were drained, and Wheelwright's teeming game decimated and dispossessed.

Had we a Wheelwright in every Australian district before it was fully settled, we would have a much clearer idea of what we have lost.

The country around Melbourne must have literarily swarmed with wild fowl . . .

I do not believe that any country in the world is better adapted by nature as a home for the water-fowl than Australia. Dreary swamps miles in extent, lagoons of immense size, where the bulrush and reed vegetate in rank luxuriance; creeks and water-holes, completely hidden from the view by dense masses of tea-tree scrub, afford unmolested shelter and breeding-places for the birds; and a few years ago, when the sound of a gun was rarely heard in the solitude of these morasses and fens, the country around Melbourne must have literarily swarmed with wild fowl.

When I first came into this country [1851] the palmy days of the duck-shooter were in their zenith; the fowls and buyers plentiful, the shooters scarce. The year previous there was not a float or big gun in this part of the colony, and the first punt that ever floated on Melbourne Swamp, was built in Melbourne Street, where the market now stands, in the morning, launched in the afternoon, fitted up with an old musket, and the birds shot and sold in Melbourne before night.

In winter, £1,000 was cleared off Melbourne Swamp and its neighbourhood by the two men who launched this punt. The diggings were then in full swing, money was like dirt, and the birds sold at any price. The buyers were not particular. Many a brace of sea-gull have been sold for 5s. and I once knew a pair of old shags with their beaks trimmed up, sold for 15s. as "rock duck". But this did not last long. The duck shooters

Black duck, *Anas superciliosa*

of that day, like the diggers, never heeded the morrow, and not one laid up for a rainy day. As the birds became scarcer, the shooters increased, and prices fell, till at the present day duck-shooting is not worth following within fifteen miles of Melbourne.

What a change has six years made in the appearance of this country. The swamps and lagoons near Sandridge are all drained or built on, and a railroad now passes over ground on which, at that day, four or five couple of ducks might be killed with ease in a night's flight-shooting . . .

Black swans, *Cygnus atratus*

H. W. Wheelwright

Explorers and other early newcomers in Australia noticed mostly three animals: the grey kangaroo, the emu and the black swan, *Cygnus atratus*.

And the swan had an edge on the other two in a way because, while they were quite uniquely Australian, the swan matched those 'at home', yet had the cussedness to be black instead of white. It reversed the proper order of things. And that summed up what many newcomers felt about Australia.

The Bushman, though he much liked a roasted young swan for Sunday dinner, had an affection for swans and their mellow bugling at night.

The Aborigines had gathered their eggs regularly, probably in very great numbers in total, and this must have had some restraining effect on numbers. In less than a century, though, tribal Aborigines in much of the country collapsed in front of the advancing white man, and one would have thought the swans had respite.

But that first European period also brought troubles for swans. Commercial eggers were active, as Wheelwright notes. Swans were often eaten and in parts, including Westernport, they were caught by running down in boats during the flightless period after breeding for their down.

By then settlements and grazing were expanding and drainage of swamps and wetlands was reducing waterfowl habitat.

But then the pendulum swung again. Irrigation schemes unknowingly provided breeding habitat for swans in their pondages and some feeding habitat on irrigated lands themselves while some of the largest concentrations of swans now to be seen in Australia are on irrigated sewage farms. From all these causes, and the swing of public sentiment, black swans are now once more abundant and secure.

. . . a graceful, elegant bird

The Black Swan is common throughout the winter after the young birds can fly, on all the large swamps and lagoons; sometimes in good-sized flocks, but generally in small companies, which I took to be the old birds and birds of the year . . .

They breed a good deal on some of the large islands in Western-port Bay, and I attribute the decrease of swans in this neighbourhood to the quantity of eggs that are yearly taken by the fishermen in this bay.

Swan-ponds near the Heads, is also a great place for them; in fact, they are by no means rare in this district, and an odd pair or so breed on most of the large swamps . . .

They are a heavy-flying bird, and don't care to rise on the wing, if they can save themselves by swimming.

The black swan is a graceful, elegant bird, not so large as the hooper at home; the shape of the beak is the same, but the cere is red, and the windpipe is not folded within the breastbone . . .

They have a very musical call-note when passing overhead on a still night; and I have listened with pleasure to the soft low notes of a pair of swans answering each other, while floating on the lagoon, by the side of which I lay at flight-time. At night they always fly low . . .

The flesh of the young swan is excellent, and one roasted in a camp oven generally with us formed the duck-shooter's Sunday dinner, whenever we could get one during the season. I wonder the skins are not more highly prized for the down, which is very thick . . .

H. W. Wheelwright

Anyone who sees a free-flying magpie goose, *Anseranas semipalmata*, in eastern Australia today—south of Queensland—has seen a rarity.

Swamp-drainage, destruction of swamp-vegetation by cattle, drought, shooting and at times deliberate poisoning virtually wiped out the species in South Australia, Victoria and most of New South Wales by early this century.

The magpie goose feeds on the roots and tubers of swamp vegetation, particularly rushes and sedges. These it digs out of sometimes quite hard mud with its strong, narrow, curved bill. It also eats seeds of water plants which it gets directly from the plants or by filtering the swamp-ooze.

With such feeding habits and with its need for substantial swamp-vegetation on which to roost and nest, the magpie goose needs large, well-vegetated swamps.

Because cattle also like swamp-vegetation, such habitat is precisely what we have not got, unless the swamps are in wildlife reserves or are fenced off by private owners who care about conservation.

The scene Wheelwright knew, of many curious necks of high-perching magpie geese sprouting above the canopy of swamp paperbark thickets on Mornington Peninsula, will never be seen there again.

So complete was the disappearance of the magpie goose from Victoria that in the 1960s the then Victorian Fisheries and Wildlife Department set up a captive breeding programme for birds brought from northern Australia to its 'Serendip' research station, north of Geelong.

After years of perfecting the techniques for successfully breeding the bird in captivity, the agency now from time to time releases batches of young magpie geese on the State's wildlife reserves. But it remains to be seen if the species can re-establish itself on a broad scale.

... the common wild goose in this district

The *Magpie*, or *Tree-goose* (ongak), is the common wild goose in this district, and, as far as I could learn, is the only common wild goose peculiar to Port Phillip.

Although met with here only in small flocks, generally I think families, there are lakes in the interior where they swarm.

I think they remained in our district throughout the year, although we used only to see them at uncertain periods, and never for long together.

As the name denotes, the colour of the magpie-goose is pied, dull black and white; it is about as large as the British brent goose, and the tail is very square.

It is a singular bird; the beak is higher in shape, and not so broad, as in the common goose, . . . the upper mandible is long, and has a powerful curve or hook . . .

The feet are semi-palmated, and formed for perching; the claws long and sharp . . .

They are generally perched high up in the tea-tree scrub, where they will sit for hours; and a curious sight it is to see them sitting upright, with their long necks stretched out on the watch.

They have a very loud, hoarse call-note when alarmed, nothing like that of the common wild goose.

The greatest curiosity of this singular bird, however, is the windpipe, which has three folds . . . but, instead of being folded within the breastbone, it lies on the left hand, outside, bedded in the flesh.

They bred sparingly with us, for I have found the nest in a thick tea-tree scrub; and I fancy the small flocks that we see in the autumn are families, which had been bred in the neighbourhood, and that they do not pack and make distant migrations like the wild geese at home.

Although a shy bird in the open, they are by no means difficult to creep up to in the thick tea-tree scrub, and many a pair I have killed right and left. They are capital eating, and will fetch from 12s. to 15s. per couple in the market . . .

Magpie geese. *Anseranas semipalmata*

H. W. Wheelwright

In four brief Australian years the Bushman learned more about our wildlife than most present Australians learn in their lives. It is sadly doubtful if one present Australian in a thousand could recognise the voice of the olive-backed oriole, *Oriolus sagittatus*, which Wheelwright called the 'green thrush'.

Certainly the Bushman had advantages. He was out in the bush most days. He was seeing a wonderfully pristine Australia before much woodland was cleared and before the worst effects of introduced pests: rabbits, feral cats, foxes and birds like starlings and sparrows began to make serious inroads.

Above all, he was *interested* and was enthused with the difference and splendour of our natural world, which most Australians are not. And so the Bushman heard and responded to the rolling advertisement of the oriole, calling its own name, 'olly-ollyole' from a treetop.

To those who know the bush, that sound, like the silvery 'falling-leaf' melody of the white-throated warbler, *Gerygone olivacea*, is still the very soul of those calm, suddenly warm, wattle-perfumed days of spring and early summer in the open woodlands of south-eastern Australia.

Olive-backed oriole, *Oriolus sagittatus*

... its wild desultory carol, borne upon the early breeze

One of the sweetest sounds in the Victorian forest, to my ear, was the loud monotonous note of the green thrush, from the topmost branch of a high gum-tree, on one of those clear delicious mornings so peculiar to the Australian spring.

Although not to be compared to the rich and varied song of the British thrush, there is a gush of melody in the few notes of the Australian bird equal to any of our finest songsters . . .

I have often and often stood at my tent about sunrise and listened to its wild desultory carol, borne upon the early breeze, laden with the fragrance of many a thousand blossoms, [and] thought how dull and senseless must that blockhead have been who described Australia as a land where the flowers have no scent and the birds no song . . .

The oriole builds a very pretty pendent nest, between two small twigs, and lays three large handsome mottled eggs; in fact, I think the nest and egg of the green thrush prettier than any I ever took in this country. It was sparingly dispersed in pairs over the whole bush, but nowhere very common . . .

H. W. Wheelwright

Southern emu-wrens, *Stipiturus malachurus*, still cling to a few isolated pockets on Mornington Peninsula. But they are going, and are certainly not part of the life-experience of the hundreds of thousands who swarm into the Peninsula on summer weekends and holidays.

A pity, because these little birds are charming, Dresden-fine creatures. Their contrary name comes from the curiously open structure of their tail-feathers.

Like the feathers of emus, these are deficient in the fine interlocking barbs and barbules that give typical feathers their close fabric. Emu-wrens have six of these tufty, thin, tail-feathers, often twice as long as the head and body. No known bird has fewer tail-plumes.

Because emu-wrens do not go far from dense, heathy cover, they are seldom seen unless you have cause to spend time in those places yourself.

Waiting for a dawn or dusk shot at duck or kangaroos, Wheelwright saw them often, and so had an experience comparatively few Australians enjoy. (Though elsewhere in coastal south-east Australia and Tasmania there are still largish, well-protected patches of habitat where emu-wrens are locally abundant.)

Southern emu-wren, *Stipiturus malachurus*

. . . scarcely larger in body than a great bumble-bee

The little *Emu* or *Pheasant Wren* was the smallest bird in our [district], scarcely larger in body than a great bumble-bee . . . The tail is about three times as long as the body, composed of six feathers (the middle ones much the longest) all clothed with fibres, after the manner of the tail-feathers of the native pheasant.

It has very small wings, and weak powers of flight,—in fact, when flying it appears to have a difficulty in bearing its long tail.

It is a busy little bird, and I liked much to watch a family of them creeping about the small scrub and heather like so many little field-mice.

We generally found them in small colonies or families, among heather, low scrub, or long grass on the plains and swamps: they were very hard to rise, and when on the wing easily knocked down with a small bush or cap. The male has a weak but pretty little song . . .

H. W. Wheelwright

Nothing shows the change that 130 years has brought to coastal south-eastern Australia better than the Bushman's recollection of the slender, long-legged ground parrot, *Pezoporus wallicus*, which he called by its alternate name, swamp parrot.

In the 1850s he undoubtedly knew the bird in the low heaths and coastal scrubs of the Carrum swamp hinterland, on Port Phillip. Today it is hard to credit that it was ever there.

The nearest ground parrot habitats now are probably the heathlands of Wilson's Promontory National Park, 150 kilometres in a direct line to the south-east of Wheelwright's old territory. And it would be gone from there, too, if the habitat were not protected by the park.

The ground parrot is one of those birds that provides a very precise gauge of environmental change. It is shaped, formed and coloured to fit its own small range of habitats and no other.

Its barred green plumage, like a tiger's stripes, has evolved to render it inconspicuous among the repeated fine stems of the rope-rush and other ground-cover it lives among. Its long legs and 'walking', rather than 'grasping' feet, fit it for moving and feeding freely on the ground. It lives on a variety of seeds of those particular plants: sedges, rushes, grasses, melaleucas, acacias, 'button-grass' (in Tasmania) and 'grass-trees'.

In most of its habitats, fire is now regarded as a necessary modifying element. Fire makes available, on a mosaic basis, vegetation of the required density and height and a continuing supply of different seeds.

In the Port Phillip region, where Wheelwright knew the ground parrot, as land-clearing and stock and human management took over, the changes to the heaths and scrubs were too rapid and too foreign to its needs. In fact, without a single ground parrot being shot, the bird would have died out or moved on, from those causes alone.

Ground parrot, *Pezoporus wallicus*

It . . . goes away very sharp before a wind

We had a curious ground parrot, common in the long grass in the plain, on the heather, and often in low tea-tree scrub (sometimes up to the knees in water) called the *Swamp Parrot* . . .

[It] is an elegant bird, both in shape and plumage; nearly as large as the rosella, but not so plump. The ground colour, light sea-green; every feather of three colours, green, black, and yellow; a long pointed tail, the feathers barred with black and yellow, and a red forehead.

The shape of the beak, head, and body, is that of the parrot. But the legs are long and bare; the claws long, straight, and pointed. In fact, it is a tree-parrot with the foot of the lark.

It lives on the ground (but I have seen them perch on the tea-tree scrub), runs much and quickly, is hard to rise, flies in jerks, goes away very sharp before a wind, and is very pretty shooting, rising from the grass and heather.

We used to find them during the whole year, frequenting different localities at different times; and although they could scarcely be said to flock, I generally rose three or four on the same spot. Dogs will set them like quail . . .

H. W. Wheelwright

Most accounts of the dingo, *Canis familiaris dingo*, by explorers or settlers speak of its depredations on stock or its presence as a semi-domesticated hunting companion of the Aborigines.

Very few discuss the dingo itself as a large predatory animal, its daily routine, or the social structure of its groups. But from his own observations virtually on Melbourne's doorstep in the 1850s, the Bushman formed a pretty accurate idea of its ways.

Belonging to a different race of the same species as the domestic dog, the dingo has a rather rigid social structure and breeds only once a year. Dogs have less structured groups and moreover can breed more than once a year. Seemingly trivial, these details are of some importance in parts of Australia like the more remote country of the Dividing Range, where feral domestic dogs interbreed with dingoes.

The dingo's natural methods of limiting its population are thus broken down and numbers of animals of mixed blood increase. This process threatens the survival of the pure-bred dingo as surely as loss of habitat caused by clearing, baits, trapping and shooting.

Yet again, Wheelwright seems to have seen the last of the natural order. Within a decade or two of his writing this, the dingo, like the tribal Aboriginal, was gone from Mornington Peninsula forever.

Dingo, *Canis familiaris dingo*

. . . the chorus was . . . very fine

The *Wild Dog*, warrigal, or dingo, is met with in all the thick forests, deeply-scrubbed gullies, in belts of timber bordering on the large plains, and in patches of tea-tree on the plains themselves . . .

Shy and retired in its habits, the wild dog is rarely seen by day, unless disturbed, lying up generally in thick patches of tea-tree scrub till evening sets in, when, like the wolf and fox of the old world, they roam abroad in search of prey . . .

[Their] colour is usually light red, but there is a beautiful variety nearly black, which is, however, rare, and, like the black fox of northern Europe, only occasionally found in a litter of red cubs.

The cry of the wild dog at night is a long dismal howl, very much resembling the horrid cry of the Swedish wolf echoing through the forests, making "night hideous'; [but] sometimes a small pack would come sweeping by our campfire at night after kangaroo, and the chorus was then very fine, when all else was still.

The wild bitch brings forth from four to six cubs, like the domestic bitch, generally in a large hollow log or old tree-root.

Unlike the wolf, they rarely hunt in large packs, and if, by chance, four or five are seen together, I fancy it is an old bitch and her cubs: . . .

[At Port Phillip] we had no lack of them on the kangaroo-ground, attracted, doubtless, by the carcasses that strewed the forests; and if ever we left a dead kangaroo out at night, it was pretty sure to be half eaten by morning . . . Their chief food appears to be kangaroo, . . . all bush animals, and offal, and birds . . .

H. W. Wheelwright

From north of Sydney to Adelaide, up to several hundred kilometres inland and in Tasmania, the eastern quoll, *Dasyurus viverrinus*, was, as the Bushman says, 'common throughout'.

Settlers in Victoria's east Gippsland wrote of quolls stretched out sunning themselves on the dog-leg log fences on frosty mornings.

Richard Howitt found nests of quolls when he felled river redgums on his property near present Alphington, a Melbourne inner suburb, in the early 1840s.

Near Paramatta in 1840 Louisa Anne Meredith said quolls were 'terribly destructive if they can get into the henhouse; not only killing to eat, but continuing to kill as many fowls or turkeys as they have time for . . .'

In the bush, quolls hunted rodents, smaller marsupials and reptiles, birds and their eggs and ate carrion. So abundant were these sources of food that one would imagine they would survive without difficulty.

But about the turn of the century, quolls suddenly became rare. No one is sure why, although there is evidence that an epidemic disease of some kind swept through the populations of several marsupials.

Coupled with loss of habitat, persecution for poultry-killing and having to survive (and compete with) foxes and feral cats, the eastern quoll population on the mainland collapsed.

In the 1940s the zoologist David Fleay tracked down a remnant in the Stony Rise country of the Victorian Western District. A small colony seems to have persisted in the Melbourne riverside suburb of Studley Park into the 1950s. But apart from those, and a few possible recent sightings, the eastern quoll now seems to be nearly, if not already, extinct on the mainland.

Fortunately in Tasmania, where the fox was never introduced, the quoll remains common.

They are common throughout the whole bush . . .

One of the commonest of all the bush animals is the little *Native Cat* or *Dasyure*, a pretty little animal, about the shape and size of a ferret; but the nose is sharper, the ears are large and pricked, and the tail is long and brushy, nearly the length of the body.

The general colour is light sandy brown, with white spots; but there is a beautiful variety, jet black spotted with white . . .

The native cat is a small beast of prey, very destructive to birds, especially poultry, and eggs.

They are common throughout the whole bush, living by day in hollow logs, old dead log fences, and holes in the ground, and at night they come out to feed on the ground; and the dogs, when hunting, generally run them up the small she oaks and honeysuckles. You rarely see a wild cat up a gum-tree.

They much frequent the belts of timber on the edges of the swamps; and I have often killed them on the beach by moonlight, coming down, no doubt, to look after the dead fish washed ashore . . .

Eastern quoll, *Dasyurus viverrinus*

John Wills

On 20 December 1860, Burke and Wills—on their bold way north to the Gulf of Carpentaria—had just left that extraordinary inland watercourse, Cooper's Creek, whose presence ever stands looming over their tragedy.

Some of the 'waterholes' in Cooper's Creek—kilometres long, two hundred or more metres wide, and up to thirty metres deep—are said never to dry up. Flanked along their banks by graceful, blue-foliaged coolabahs and river redgums of immemorial age and unchallenged girth—great knobbly silver and rose-trunked trees with enormous limbs—these Cooper waterholes are splendid to come upon after driving across windswept upland gibber plains or up the Strzelecki Track.

They are also highway and refuge to thousands of screaming little corellas, fleets of pelicans, dark friezes of cormorants, ducks, teal, herons, egrets, even silver gulls from the distant sea.

Birds of prey: whistling kites, black kites, little and peregrine falcons, soar high against a sky of peculiar luminescence. It is all so different from what one 'knows' to be the far inland. How much more wonderful was it for explorers like Sturt, who discovered it in 1845 or Burke and Wills, who paused there on their historic way north to the Gulf in December 1860 and died there on the way back, stranded by the ineptitude of their organising committee.

The following are excerpts from the diary of John Wills, descriptive of the surroundings and wildlife on the Cooper, as they made their way along the creek and then headed north along Eyre Creek in December 1860 and January 1861—at the baking height of the inland summer. It is possible that the fish were golden perch, *Macquaria ambigua*, then common in Cooper's Creek and other inland waters.

. . . they might have been fried in their own fat

Wednesday, Dec. 19 Started at a quarter past eight a. m. Leaving what seemed to be the end of Cooper's Creek we took a course a little to the north of west, intending to try and obtain water in some of the creeks that Sturt mentioned that he had crossed . . .

At fifteen miles, we halted where two large plains joined. Our attention had been attracted by some red-breasted cockatoos, pigeons, a crow, and several other birds, whose presence made us feel sure that there was water not far off; but our hopes were soon destroyed by finding a claypan just drying up. It contained just sufficient liquid to make the clay boggy . . .

Thursday, Dec. 21 [Wills makes an error in the date] . . . Our course . . . took us through some pretty country, lightly timbered and well grassed. We could see the line of creek timber winding through the valley on our left . . . At two miles further we came in sight of a large lagoon . . . and at three miles more we camped on what would seem the same creek as last night, near where it enters the lagoon.

The latter is of great extent, and contains a large quantity of water, which swarms with wildfowl of every description . . .

There was a large camp of not less than forty or fifty blacks near where we stopped. They brought us presents of fish, for which we gave them some beads and matches.

These fish we found to be a most valuable addition of our rations. They were of the same kind as we found elsewhere, but finer, being nine to ten inches long, and two or three inches deep, and in such good condition that they might have been fried in their own fat . . .

Golden perch, *Macquaria ambigua*

Chequered swallowtail, *Papilio demolens*

Ernest Giles

Mount Olga in the south-west Northern Territory was seen and named by Ernest Giles, that tough, sardonic desert explorer, toward the end of 1872. Travelling through baking and often burned-out semi-desert scrubs south of Mt Udor, he saw it looming far to the south.

It promised a change in his present featureless surroundings, and as Giles knew better than most, large hills often meant water. But the approach was eventually blocked by the bottomless, gluey blue mud of dry, salty Lake Amadeus, and Giles disgustedly turned back.

On his second expedition, which left Ross's Waterhole on the Alberga River (about 500 kilometres south of Alice Springs) in August 1873, Giles travelled west through the Musgrave Ranges, then turned north through a break toward the western end of those ancient, rounded hills. He came up to the Olgas from the south, through open sandhills of 'spinifex' and desert oaks, then through stony, scrubby country, where he camped without water or feed for his horses.

Nothing broke the enormous silent presence of those stupendous red walls and domes. Only the weak 'chickowee' of a grey-headed honey-eater, elfin against that overpowering backdrop and the mute wagon-tracks of the South Australian explorer Gosse, who had been to the Olga since Giles named it the year before.

Giles's description of Mt Olga, before the European era overtook the Centre, while it was still part of the Dreamtime, is fascinating.

To heighten the hugeness of the thing, he chose, perhaps unconsciously, to emphasise a butterfly. This was the chequered swallowtail, *Papilio demolens*, a lovely thing widespread in Asia and the Australian region. This excerpt is from Giles's *Australia Twice Traversed: The Romance of Exploration . . .*, 2 vols (London, 1889).

Time, the old, dim magician, has ineffectually laboured here . . .

The appearance of this mountain is marvellous in the extreme, and baffles an accurate description . . . it is formed of several vast and solid, huge, and rounded blocks of bare red conglomerate stones, being composed of untold masses of rounded stones of all kinds and sizes, mixed like plums in a pudding, and set in vast and rounded shapes upon the ground.

Water was running from the base, down a stony channel, filling several rocky basins. The water disappeared in the sandy bed of the creek, where the solid rock ended . . .

I made an attempt to climb a portion of this singular mound, but the sides were too perpendicular; I could only get up about 800 to 900 feet, on the front or lesser mound; but without kites and ropes, or projectiles, or wings, or balloons, the main summit is unscaleable. The quandong fruit here was splendid—we dried a quantity in the sun.

Some very beautiful black and gold butterflies, with very large wings, were seen here and collected. The thermometer to-day was 95° in the shade. We enjoyed a most luxurious bath in the rocky basins. We moved the camp to softer ground, where there was a well-grassed flat a mile and a half away . . .

The appearance of Mount Olga from this camp is truly wonderful; it displayed to our astonished eyes rounded minarets, giant cupolas, and monstrous domes. There they have stood as huge memorials of the ancient times of earth, for ages, countless eons of ages, since its creation first had birth. The rocks are smoothed with the attrition of the alchemy of years.

Time, the old, dim magician, has ineffectually laboured here, although with all the powers of ocean at his command; Mount Olga has remained as it was born; doubtless by the agency of submarine commotion of former days, beyond even the epoch of far-back history's phantom dream. From this encampment I can only liken Mount Olga to several enormous rotund or rather elliptical shapes of rouge mange, which had been placed beside one another by some extraordinary freak or convulsion of Nature . . .

Ernest Giles

Leaving Mt Olga, Ernest Giles and his two companions travelled south-west. Then in October and November 1873, with the pitiless inland summer advancing, they pushed on and on west in a vain attempt to cross to the Western Australian coast. Desert defeated him. Giles back-tracked, moved north, tried again. This time Giles very nearly left his bones in the desert.

At the lowest point in his ordeal, Giles's life was saved by finding a baby wallaby left by its mother. He seized it and ate it whole.

Putting themselves to such trials, the explorers were often obliged to live off the land. So even before his desert nightmare Giles had a healthy enthusiasm for fresh wild food. It helps explain his delight at finding several active incubation mounds of the malleefowl or lowan, *Leipoa ocellata*, in mallee country west of the remote Tompkinson Ranges, just over the border into Western Australia.

The fragility of malleefowl egg-shells seems to be an adaptive device to enable the chick to fracture the shell from within, while buried under perhaps 60 centimetres of sand, humus and small sticks. Unassisted by its parents, it then digs its way by instinct to the surface and totters off to make its lone way in the world.

But malleefowl are now mostly gone from the far-inland. The reason for this collapse can probably be found in the extraordinarily fine tune of their relations with their environment. They subsist mostly without surface water, satisfying their needs for moisture (including that for the sixteen or more large eggs the females produce each year) from insects, dew and from small green plants.

When rabbits, goats, sheep and cattle move into malleefowl habitat, they can remove enough of the small plants to tip the fine balance of survival against the bird.

These birds . . . build extraordinarily large nests of sand . . .

The country was very peculiar . . . it was open, covered with tall triodia, and consisted of almost entirely of limestone. At intervals, eucalyptus-trees of the mallee kind, and a few of the pretty-looking blood-wood trees and some native poplars were seen; . . .

To-day we came upon three lowans' or native pheasants' nests. These birds, which somewhat resemble guinea-fowl in appearance, build extraordinarily large nests of sand, in which they deposit small sticks and leaves; here the female lays about a dozen eggs, the decomposition of the vegetable matter providing the warmth necessary to hatch them.

These nests are found only in thick scrubs. I have known them five to six feet high, of a circular conical shape, and a hundred feet round the base. The first, though of enormous size, produced only two eggs; the second, four, and the third, six. We thanked Providence for supplying us with such luxuries in such a wilderness.

There are much easier feats to perform than the carrying of lowans' eggs, and for the benefit of any readers who don't know what those eggs are like, I may mention that they are larger than a goose egg, and of a more delicious flavour than any other egg in the world. Their shell is . . . so terribly fragile that, if a person is not careful in lifting them, the fingers will crunch through the tinted shell in an instant. Therefore, carrying a dozen of such eggs is no easy matter.

I took upon myself the responsibility of bringing our prize safe into camp, and I accomplished the task by packing them in grass, tied up in a handkerchief, and slung round my neck . . .

Malleefowl, *Leipoa ocellata*, at nest-mound

Malleefowl, *Leipoa ocellata*, at nest-mound.

Lotusbird, *Irediparra gallinacea*, and young

Carl Lumholtz

Carl Lumholtz (1851–1922) was one of those remarkable young, educated Europeans of the 19th century who came to Australia in search of knowledge and adventure.

Abandoning theology at the University of Norway, Lumholtz had turned to anthropology, zoology, animal collecting and writing.

From Lumholtz's four years in Australia (1880–84) came his first book, *Among Cannibals* (London, 1889), a vivid account of his life in the Queensland wilds.

Now a classic, it gives a timeless picture of wild north-eastern Australia as it remained up to a century ago, a picture caught before that life was swept away by European settlement.

Arriving at Adelaide in August 1880, Lumholtz visited Melbourne, Sydney and Brisbane, then took ship to Rockhampton to stay at Gracemere Station, the famous property of the Archer family, where he remained until August 1881. Gracemere, then as now, had a splendid sub-tropical lagoon, with a haze of blue water-lily flowers across its surface.

Here, Lumholtz met that most fascinating of swamp-dwellers, the lotusbird, *Irediparra gallinacea*, whose extremely long, slender toes support it on the leaves of water-lilies. This floating vegetation is the lotusbird's world. Here it feeds on insects and seeds and here it makes its nest, a sodden floating mass of weeds which barely supports the handsome eggs.

This is an exposed habitat, and calls for special survival techniques. Among other remarkable attributes, the lotusbird has been shown to deliberately carry its downy young from danger by crouching and allowing them to clamber up under its wings. When the young are larger they dive to escape danger, as Lumholtz observed.

The most striking bird on the lagoon . . .

The most striking bird on the lagoon is doubtless the beautiful . . . lotus-bird . . . [which] sits on leaves that float on the water, particularly those of the water-lily.

Blue water-lilies are found in great numbers along the edge of the lagoon, and hence the lotus-bird is very common here. It is somewhat larger than a thrush, and has very long legs, and particularly highly developed toes, which enable it to walk about on the floating leaves. Its food consists chiefly of snails and insects, which it usually finds by turning the lily leaf. Its simple nest is also built on the leaves.

The eggs, which are a beautiful brown with lines and spots, are considered very rare, and are remarkable both on account of their form and colour . . . The young look very funny on account of their long legs and big toes as compared with their small bodies.

The grown bird is not shy, but the young are extremely timid. I had once or twice seen the old birds with young, but as soon as I approached them, the young always disappeared, while the old birds walked about fearlessly, as if there was no danger. It long remained a mystery to me, how they could conceal themselves so well and so long, but one day the problem was solved.

An old bird came walking with two young ones near shore. I hid behind a tree and let them come close to me. As I suddenly made my appearance, the small ones dived under the water and held themselves fast to the bottom, . . .

Carl Lumholtz

In May 1882, Carl Lumholtz sailed north, and by July had established himself at Herbert Vale, an all-but-deserted property some 50 kilometres up the magnificent Herbert River valley from Ingham in north-east Queensland.

Here he made acquaintance with the still little-civilised Aboriginal people of the region, people very different from Aborigines elsewhere in Australia.

One of his objects was to study the natives 'in their original condition . . . in their actual conditions of existence . . .' To do this he learned to live like a native and, as he said, frequently walked all day without eating.

His supplies consisted usually of a dozen pieces of salt beef in a bag, about twelve kilograms of wheat flour for damper, and a small sack of sugar. Apart from these, Lumholtz lived on what he could shoot, finding safety in the magical power of his gun and that of the 'baby gun' that never left his belt. For 'money', a persuader for the natives to assist him, he carried a supply of clay pipes and tobacco, of which the natives were very fond, it being their only stimulant.

Thus equipped, Lumholtz travelled and lived at intervals for nearly a year, going out as far as 150 kilometres from his base, often solely in the company of Aborigines who had never seen or spoken with another white man.

In this way he learned their customs and met the best hunters among them. He also learned a great deal about the wildlife and collected several animals not then known to science.

His meeting with the Australian cassowary, *Casuarius casuarius*, is a good example of his method of work in this country still in its cloak of ages. In this account, 'scrub' should be read as rainforest.

. . . the stateliest bird of Australia

It is not easy to penetrate this scrub, which is so dense that one has scarcely elbow-room; but along the rivers there is more breathing space. Here beautiful landscapes are often disclosed to view; the most varied trees vie with each other for a place along the quiet stream; while creeping and twining plants hang in beautiful festoons over the water.

On first entering the scrub, the solemn quiet and solitude which reign there are striking. You work your way through it by the sweat of your brow; you startle a bird, which at once disappears, and your prevailing impression is that there is no life. But if you come there in the early morning or towards evening, and sit down quietly, it is surprising to see the birds approaching gently, as if they had been called, and disappearing as noiselessly as they came . . .

One of the first birds you notice is the cat-bird . . . which makes its appearance towards evening, and has a voice strikingly like the mewing of a cat. The elegant metallic-looking "glossy starlings" . . . greedily swoop with a horrible shriek upon the fruit of the Australian cardamon tree. The ingenious nests of this bird were found in the scrubs near Herbert Vale—a great many in the same tree. Although this bird is a starling, the colonists call it "weaver-bird."

In these scrubs the proud cassowary, the stateliest bird of Australia, is also to be found.

I had already made several vain attempts to secure a specimen of this beautiful and comparatively rare creature. We had frequently seen traces of it under the large fig-trees, the fruit of which it eats . . . [but] often approached without seeing it, for it is exceedingly shy and departs on the slightest noise, consequently it is very difficult to get a shot.

On October 6 the natives brought me two eggs and a young bird just hatched. I at once requested one of them to guide me to the nest, whither I took it, hoping thereby to attract the old bird.

Near the nest, which was formed of a not very soft bed of loose leaves massed together, we placed the young one and then stepped aside to see what would happen. It first began to run after us, but as soon as it lost sight of us, commenced to cry violently.

After a lapse of about ten minutes we suddenly heard the voice of the cassowary, which usually sounds like thunder in the distance, but now, when calling its young, it reminded us of the lowing of a cow to its calf. The sound came nearer, and soon the beautiful blue and red neck of the bird appeared among the trees, and its black body became visible. It stopped and scanned its surroundings carefully in the dense scrub, but a charge of No. 3 shot, fired from a distance of fifteen paces, laid it low.

My black companion gave a shout of victory, and ran back to the camp to get some men to carry the precious burden home. Six natives took turns in carrying it to the station, where I at once set to work skinning it. The blacks made a feast of its flesh, and the skin formed a valuable addition to my collection.

It was an unusually fine specimen of a male, who thus appears to care for the young, at least in the early stage . . .

Cassowary, *Casuarius casuarius*, and young

Carl Lumholtz

At Herbert Vale, Lumholtz made the acquaintance of Willy, a local Aboriginal who assured him that on 'his land', further up into the highlands surrounding the Herbert River Gorge, were several very special animals at that stage known to Lumholtz only as names like 'toollah' and 'boongary'.

Lumholtz went along and found himself on a spectacular and exhausting journey which took him into maddening thickets of lawyer-palm, *Calamus muelleri*.

The effort proved worthwhile. The animal they collected toward evening on the first day high up on the range proved to be the 'toollah'—a ringtail possum new to science. Not only that, but it was one of the handful of mammals on earth to have greenish fur: actually a visually-deceiving combination of yellow, green, grey and white fur.

At Lumholtz's university in Norway the animal was later named *Pseudocheirus archeri*—the first name for the genus to which it seemed to belong, the second for the Archer family of Gracemere.

Today the green ringtail possum is known to inhabit highland rainforests above 250 metres between Townsville and Cooktown in northeast Queensland, and nowhere else. It is one of the few possums not to use tree hollows, nor does it build a nest, or 'drey', like that of the common ringtail.

Clothed in fur sufficiently dense to keep it warm and to shed tropical deluges, it sleeps solitary by day on an open branch, upright in a tight ball. It sometimes also feeds by day, eating mostly the leaves of rainforest trees, notably figs. So the greenishness of its fur is clearly an adaptation to an exposed way of life on branches that, as Lumholtz observed, are often padded with green moss and lichens.

The kingfisher Lumholtz saw on his trek up the river was the azure, *Ceyx azurea*.

. . . this animal . . . looks very much like a moss-grown tree-trunk

The next day we made our ascent along the river. We had to wade most of the time. The natives made the most remarkable progress, stepping lightly on the stones, while I with my shoes on could scarcely keep pace with them. It was a long and difficult road to travel.

Weary and thirsty, I often stooped to drink the cool water, and to bathe my head in it. But I was cheered by the sight of the luxuriant and beautiful surroundings. Trees and bushes formed a wall along the mountain stream, overhanging the babbling water.

In the woods all was dark and damp, but on gazing upward I saw the tree-tops flooded with the most brilliant sunlight, which occasionally penetrated through the branches, and above us was spread the sky in an infinite expanse . . .

Now and then we startle from its branch the beautiful little indigo blue and red kingfisher . . . which with quick wing-strokes flies before us up the stream. Among the tree-tops the large brilliant blue or green butterflies . . . flutter . . .

As we ascend, the landscape gradually grows wilder and more picturesque. The river gorge becomes narrower, the amount of water diminishes, and no more kingfishers are seen.

The palms are replaced by gigantic tree-ferns, which here, in the damp rocky clefts, spread their mighty leaves in all their splendour over trickling brooks, which frequently disappear in little waterfalls down steep precipices. . . .

On the summit we . . . meet with scenes of a wholly different character. Here is the real home of the lawyer-palm, which grows on small hills, where the soil consists of a deep black mould, and consequently is so fertile that it produces everything in the greatest abundance.

Progress is difficult here, because this palm grows into immense heaps twenty to twenty-eight feet high, one by the side of the other, and often firmly woven together. In this way large connected masses are formed, appearing like an impenetrable wall. But the native usually finds a narrow passage, through which he can crawl, but not without getting badly scratched . . .

Azure kingfisher, *Ceyx azurea*

Green ringtail, *Pseudocheirus archeri*

Working our way up the side of the mountain near the summit, the natives called my attention to an animal the size of a cat, which ran about in the branches of a tree. They called it toollah.

It was late in the afternoon when I killed this animal, which proved to be a kind of opossum now known in zoology by the name of *Pseudocheirus archeri*; it has a peculiar greenish-yellow colour with a few indistinct stripes of black or white, and thus looks very much like a moss-grown tree-trunk.

Though it is a night animal, it also comes out about three or four o'clock in the afternoon, and is the only one of the family which appears in the daytime . . .

119

Carl Lumholtz

Lumholtz often learned the Aboriginal names of animals before he met them and frequently the identity behind the name was a complete surprise. Such was the case with the 'boongary', a celebrated creature described to him with enthusiasm from the time he arrived.

Relaxing in a native camp at the end of a hot day, he was suddenly presented with the animal. It proved to be new to science and is to this day known as Lumholtz's tree-kangaroo, *Dendrolagus lumholtzi*.

Australian marsupials are thought to be descended from smallish, possum-like ancestors. Isolated while the continent drifted north from present Antarctica towards Asia over some 20 million years, our marsupials radiated into an astonishing array of forms to match the opportunities offered by enormous changes in Australian climates and vegetation over that period.

Up to the mid-Miocene, some 15 million years ago, climates were warm and humid and rainforests covered much of the continent. To exploit such habitat there arose a suite of possums, including several large ones, the cuscuses, and a small group of wallabies that perhaps learned to hop up sloping trunks, as some rock wallabies still do.

Those new tree-kangaroos-in-the-making had to carry out many adjustments to suit the new way of life: forelegs became longer, hindlegs shorter; 'feet' were equipped with soft, granulated pads for gripping the branches. Tails, unable to revert and become prehensile, developed into loose pendulums that could arch up to become long balancing poles. Teeth became adapted for leaf and fruit eating. Even the lay of the fur adjusted itself to suit a way of life crouched on high branches: it falls outward from a whorl on the shoulders, so that a tropical deluge is shed more effectively.

. . . the most beautiful mammal . . . in Australia

I had just eaten my dinner, and was enjoying the shade in my hut, while my men were lying round about smoking their pipes, when there was suddenly heard a shout from the camp of the natives.

My companions rose, turned their faces toward the mountain, and shouted, *Boongary, boongary*! A few black men were seen coming out of the woods and down the green slope as fast as their legs could carry them. One of them had a large dark animal on his back.

Was it truly a boongary? I soon caught sight of the dog "Balnglan" running in advance and followed by Nilgora, a tall powerful man.

The dark animal was thrown on the ground at my feet, but none of the blacks spoke a word. They simply stood waiting for presents from me.

At last, then, I had a boongary, which I had been seeking so long. It is not necessary to describe my joy at having this animal, hitherto a stranger to science, at my feet . . . I at once began to skin the animal . . .

I at once saw that it was a tree-kangaroo (*Dendrolagus*). It was very large, but still I had expected to find a larger animal, for according to the statements of the natives, a full-grown specimen was larger than a wallaby—that is to say, about the size of a sheep. This one proved to be a young male.

The tree-kangaroo is without comparison a better proportioned animal than the common kangaroo. The fore-feet, which are nearly as perfectly developed as the hind-feet, have large crooked claws, while the hind-feet are somewhat like those of a kangaroo, though not so powerful. The sole of the foot is somewhat broader and more elastic, on account of a thick layer of fat under the skin. In soft ground its footprints are very similar to those of a child . . .

Upon the whole, the boongary is the most beautiful mammal I have seen in Australia. It is a marsupial, and goes out only in the night. During the day it sleeps in the trees, and feeds on the leaves. It is able to jump down from a great height and can run fast on the ground.

So far as my observation goes, it seems to live exclusively in *one* very lofty kind of tree, which is very common on the Coast Mountains, but of which I do not know the name. During rainy weather the boongary prefers the young low trees, and always frequents the most rocky and inaccessible localities. It always stays near the summit of the mountains, and frequently far from water, and hence the natives assured me that it never went down to drink.

During the hot season it is much bothered with flies, and then, in accordance with statements made to me by the savages, it is discovered by the sound of the blow by which it kills the fly. In the night, they say, the boongary can be heard walking in the trees . . .

Lumholtz's tree-kangaroo, *Dendrolagus lumholtzi*

Carpet python, *Morelia variegata*

Carl Lumholtz

Probably nothing in *Among Cannibals* demonstrates Carl Lumholtz's adoption of the native way of life in the north Queensland rainforests better than this spectacular account of the hunting of carpet pythons, *Morelia variegata*, high in arboreal clumps of bird's nest or elkhorn ferns.

Can you imagine a newly-arrived European (or Australian) zoologist being so delighted with baked python-liver?

After a bare three years in Australia here he was, living and seeing wildlife much as the Aborigines had done for 50 000 years, as few modern white men have ever done. It is an extraordinary story and, one must believe, wholly true.

During this season the natives are . . . occupied in hunting snakes

Winter had now set in in earnest . . . a more agreeable temperature than Northern Queensland during this season of the year can scarcely be conceived, especially toward sunset. . . .

During the night so much dew falls that the woollen blanket becomes saturated if one sleeps beneath the open sky. Walking in the grass in the morning is almost like wading in a river. One becomes drenched to the hips. But what glorious mornings! They stimulate a person to work, and their freshness awakens all the joys of life. . . .

During this season the natives are much occupied in hunting snakes, which during the winter are very sluggish . . . The blacks are particularly fond of eating snakes, but they do not, like many of the southern tribes, eat poisonous serpents.

One of the snakes most commonly eaten is the [carpet python]. During winter it seems to prefer staying in the large clusters of ferns found on the trunks of trees. At night it seeks shelter from the cold among the leaves, but during the daytime it likes to bask in the sunshine, which enables the natives to discover and kill it with their clubs. If attacked it may bite with its many and sharp teeth, but the wound produced is not dangerous.

These ferns grow in wreaths round the large trunks of trees, and look like the topsails of a ship . . . and like the orchids, which grow pretty much in the same manner, are constant objects of interest to the natives, for in them they find not only snakes, but also rats and other small mammals. . . . They therefore, as a rule, take the trouble to climb the tree to make the necessary search. They discover the snakes at a great distance, though the wreath may be fifty to sixty yards above the ground.

We were at one time travelling along one of the mountain streams, while the blacks as usual kept a sharp look-out and examined the numerous clusters of fern in the scrub. Suddenly they discovered something lying on the edge of one of these fern clusters, but very high in the air. Notwithstanding their keen eyesight, they were unable to make out whether it was a serpent or a broken branch, so a young boy, whom I usually called Willy, climbed up in a neighbouring tree to investigate the matter.

Ere long he called down to us, *Vindcheh! vindcheh!*—that is Snake! snake! I was very much surprised, for the object looked to me like an old leafless limb of a tree. Willy came down at once, and lost no time in ascending the tree where the serpent was lying.

When he had obtained a foothold near the fern wreath, he broke off a large branch and began striking the serpent, which now showed signs of life. The lazy snake soon received so many blows on the head that if fell down, and proved to be more than ten feet long. While we were taking a look at it we heard Willy, whom it was almost impossible to discover so high up in the tree, call down that he had found another snake, and this made the blacks jubilant.

. . . As quickly as possible the camp fire was made and stones were heated; for snakes are one of those delicacies which are prepared in the most *recherché* manner. The snakes were first laid carefully in circular form, in order that they might occupy as small a space as possible; each forming a disc fastened together with a reed, they looked like the rope-coils made by sailors on the deck of a ship. . . .

. . . Snake-flesh has a white colour, and does not look unappetising, but it is dry and almost tasteless. The liver, which I found excellent, tastes remarkably like game, and reminds one of the best parts of the ptarmigan.

Carl Lumholtz

Carl Lumholtz left north Queensland in 1884 after nearly a year spent mostly living and hunting with the Aborigines. He had experienced an Australia that few educated Europeans ever experienced, at least in the same degree.

It is clear from his words on leaving Herbert Vale that the spirit of those splendid rainforests would stay with him. Perhaps the epitome of those forests, the very condensation of their beauty and the essence of their wildness, is Victoria's riflebird, *Ptiloris victoriae*, whose male is a splendid creature of velvet black plumage with shimmering, iridescent crown, throat and upperbreast.

This vision is seen most spectacularly in the territorial display when in the morning on a stump or low branch, he performs noisily, expanding his glossy, shot plumage in the sunlight, spreading his wings and raising wide his open bill, exposing its bright, yellow-green lining to any passing females.

. . . there is indeed melancholy, but also untold beauty

Upon the whole, I took leave of the country of the blacks and my interesting life in the mountains with strange feelings in my breast.

Some of the impressions derived from this grand phase of nature I shall never forget. When the tropical sun with its bright dazzling rays rises in the early morning above the dewy trees of the scrub, when the Australian bird of paradise arranges its magnificent plumage in the first sunbeams, and when all nature awakens to a new life which can be conceived but cannot be described, it makes one sorry to be alone to admire all this beauty.

Or when the full moon throws her pale light over the scrub-clad tops of the mountains and over the vast plains below, while the breezes play gently with the leaves of the palm-tree, and when the mystic voices of the night birds ring out on the still quiet night, there is indeed melancholy, but also untold beauty . . .

Victoria's riflebird, *Ptiloris victoriae*

Knut Dahl

Born in 1871, the Norwegian naturalist Dr Knut Dahl spent much of the period 1894–96 in northern Australia investigating and collecting wildlife before returning to Norway to follow an academic career. Part of that period was spent in Arnhem Land, part in the Kimberleys. Dahl recounted his Australian experiences in his book, *In Savage Australia* (London, 1926).

Among other noteworthy finds, Dahl discovered the very localised black-banded pigeon near the South Alligator River, and collected the first specimens. He also collected the type specimens of the wogoit or rock ringtail possum, the chestnut-quilled rock-pigeon and the hooded parrot.

Dahl had a lively and meticulous eye for detail and a freshness of expression which carries through translation. His description of the little rock wallaby, *Peradorcas concinna*, one of the smallest of its tribe and of a sugar glider, *Petaurus breviceps* which nightly raided their saddlebags for sweet food, are splendid examples of a delighted, observant man coming freshly upon the living treasure of a new world.

At the time of writing he was camped near Burrundie, west of the headwaters of the Mary River in the Northern Territory.

Sugar glider, *Petaurus breviceps*

The swiftness and agility of the animal are almost inconceivable . . .

The beautiful little rock wallaby . . . occurred in enormous numbers. This species, by the natives named 'balwak', I have practically not met with outside this peculiar granite formation. In the deep chambers and crannies of the enormous piles of boulders, where the rays of sun never penetrate, the little balwak spends the day lightly sleeping. It is very easily flushed and runs from rock to rock with astonishing activity.

The swiftness and agility of the animal are almost inconceivable, and, when observing one of these small wallabies running in the open at top speed, one might also believe it to be the shadow of some bird flying swiftly overhead.

At sundown they come out, and mounting high stones and rocks before commencing to feed they appear for some time to enjoy the cool evening breeze and the glow of the tropical sunset. The most minute noise will then alarm them, and silently, like flitting shadows, their light forms will disappear among the broken boulders of the granite.

Occasionally they approach water in order to drink but they do not appear to be so dependent on water as many of the other kangaroo species. They breed all the year round. Only one young is born at a time, and the mother leaves it immediately when in danger. If one of these wallabies be wounded, she will instantly pull the young out from her pouch, flinging it aside, possibly to be able to run with greater ease . . .

The little flying squirrel was also resident in the surrounding forest. Almost every morning these curious animals paid our camp a visit just before the break of day.

They rummaged in our saddle gear and in our provisions and pack bags. But as soon as we moved to pick up a gun they would run swiftly up a large tree just in front of our awning and spreading their parachute would sail into the dusk, disappearing before we could shoot . . .

Knut Dahl

While staying at Arenbarra cattle station near the mouth of the Adelaide River in mid-1894, Knut Dahl experienced the spectacular dry season concentration of water-birds, especially of the Australian crane, the brolga, *Grus rubicundus*.

To a naturalist newly arrived in Australia, accustomed to seeing European cranes only in small numbers, the spectacle was memorable.

This dry-season concentration of brolgas on the coastal plains of northern Australia is part of an annual routine. During the wet summer season, brolgas disperse in pairs and breed in the extensive shallow swamps and lagoons that spread across the coastal plains.

Food is usually plentiful. The one or two young are fed by the parents and remain dependent on them for up to a year.

As the water-level of the swamps falls in April and May with the onset of the dry, the breeding territories break up and brolgas of all ages gather on the remaining shallow waters where food is concentrated.

In these gatherings, young birds, unmated adults forming pairs and mated pairs cementing pair-bonds indulge in the 'dancing displays' which have made the cranes famous.

As the dry season advances and the remaining shallow waters dry out, the brolgas must feed for long periods each day, digging in the dry mud for the tubers of sedges and other plant-food. This stressful period continues until the first rains of the next wet season soften the soil and begin to fill the coastal wetlands, setting the cycle off once more.

. . . I had never imagined . . . such flocks

In the enormous plains one met a bird that really was an important feature in the landscape. This was the Australian crane . . . or 'native companion', as it is commonly called. I had certainly observed this beautiful species on the Daly and also shot one specimen; but I had never imagined that such flocks as I saw in these plains on the Adelaide really existed. When riding about during the first days of my stay at Arenbarra I became aware of enormous blue-grey strips and patches in the distant horizons of the plains. At first I took these to be lagoons and waters. Approaching, however, I perceived them to be birds, the cranes which in flocks of hundreds covered the flats.

They would not allow me to get nearer than about three hundred yards, but this was close enough for me to observe their queer and grotesque movements. Those birds who were not feeding occupied themselves after the fashion of cranes, in dancing. They lifted their wings and stretching their heads forward with curved necks would jump about, indulging in the mad and burlesque steps peculiar to the crane's dance. New arrivals were constantly coming from all parts of the plain. The air was full of them, and their clear and high-pitched trumpet note resounded everywhere.

Brolgas, *Grus rubicundus*

When they rise they follow the tactics of all long-winged birds. They spread their wings and, running swiftly forward, jump. Reaching earth again, they continue the run, jump again, and finally with slowly moving wings glide upwards at a very moderate angle.

If a bird intends to go far it screws itself gradually upwards in immense circles. When it wants to alight again its tactics are also interesting.

It suddenly lets the legs drop, the neck is drawn back and the tail is depressed. Immediately the bird hurtles down through the air, using the half-open and, as it were, hanging wings as a brake. This terrific speed lasts until the bird is a few yards distant from the earth. Then the wings are fully spread again. Tail and neck are straightened once more, and, stretching the legs forward, the bird skims along the earth until the speed subsides. Finally it lands, jumping high in the air on the first impact, and settling with inimitable grace sounds its clear trumpet-note.

As a fact these cranes fly very high, and I am certain that I have seen them drop from an altitude of six to eight hundred feet at an angle of about 45 degrees . . .

K. H. Bennett

The coming of Europeans to Australia had opposing effects on the wedge-tailed eagle, *Aquila audax.*

First, it deprived the eagle of much of its natural prey of small marsupials. The now classic combination of the spread of stock, over-grazing and drought coupled with the advance of introduced rabbits, foxes, cats and goats, wiped out or at least decimated populations of no less than fifteen small to medium-sized mammals in eastern and inland Australia.

But we quite unintentionally compensated for this devastation of the wedge-tail's food-base by introducing the European rabbit, *Oryctolagus cuniculus*, which provided eagles and other birds of prey with a bountiful, easily-caught and delicious staple.

These days, despite many more observers in the field, reports of successful attacks by wedge-tailed eagles on dingoes, kangaroos or koalas in south-eastern Australia and Tasmania are rare.

This seems to be because wedge-tails now have no need to tackle such big game. We have provided easier substitutes: rabbits, carrion in the form of dead lambs and sheep, or kangaroos killed along outback roads by spotlight shooters or vehicles.

So the rather spectacular observations by K. H. Bennett of hunting wedge-tailed eagles in Gippsland, Victoria, although made when rabbits were already on the rise, reflect old habits that were even then in process of change.

Bennett was later a pastoralist of the Mossgiel district of inland New South Wales. A keen and accurate observer of wildlife, he corresponded with leading ornithologists and many of his notes, including those here, were quoted by A. J. North in his great *Nests and Eggs of Australian Birds* (Sydney, 1901–04).

Wedge-tailed eagle, *Aquila audax*

. . . they do not always confine themselves to small quarry

Although Wedge-tailed Eagles prey to a large extent on rabbits, they do not always confine themselves to small quarry of this description, for in Gippsland I have often seen them attack full grown native bears, and on one occasion a pair attacked a half grown kangaroo, but I did not see the result, as it was in a thickly timbered place.

When they passed me the kangaroo was going at its utmost speed, with one Eagle perched on its neck and flapping its wings about its face, evidently with the intention of terrifying and confusing it, the other Eagle flying alongside.

I also saw, on another occasion, a pair kill a full grown dingo. I did not see the commencement of the attack, but when I came across them they had evidently been at the dog some time, for he was very much exhausted, and was staggering along in an aimless manner. One Eagle was perched on the dog's neck and flapping its wings, the other perched on his loins; occasionally the latter would turn his head and snap in a feeble manner at the Eagle, who would simply fly up, and the next instant drop on the loins again. This continued for some time, the dingo evidently getting weaker and weaker, until he stumbled, fell and lay perfectly still.

I saw the Eagles walking round him, and then begin tearing at his flank with their bills. I waited and watched for some little time longer, and then rode up and found the dingo, which was in fine condition, quite dead . . .

Although the Bridled Wallaby . . . is not strictly nocturnal on the Lower Lachlan River, in Southern New South Wales, it is very rarely met with away from the shelter of the dense bush or scrub during the day. The reason of this is its dread of its terrible enemy, the Wedge-tailed Eagle, this bird destroying great numbers, particularly during the nesting season, when the nests and the ground beneath are strewed with the remains of this animal.

Koalas, *Phascolarctos cinereus*

Torres Strait pigeon, *Ducula spilorrhoa*

E. J. Banfield

E. J. Banfield, 'the Beachcomber', born in England in 1852, came to Australia as a child, later becoming a journalist.

It was while on a newspaper in Townsville in the 1890s that the combination of editorial worries, uncertain health and the lure of the splendid tropical coastline away to the north, persuaded Banfield and his devoted wife to alter their lives.

They secured a lease, ordered a prefabricated cedar hut which could be dismantled if the experiment failed, and embarked for Dunk Island, four kilometres off South Mission Beach, midway between present Townsville and Cairns.

On Dunk the Beachcomber, as Banfield became known, revived wonderfully in health, worked, built and gardened like a demon and lived on happily until 1923. He also found time to write several charming books about their Robinson Crusoe existence which made his name known far beyond Australia.

The following excerpts are taken from *The Confessions of a Beach-comber* (London, 1908). The first presents the splendid Torres Strait pigeons, *Ducula spilorrhoa*, coming to nest in their multitudes.

This large white fruit-pigeon winters in New Guinea, the Aru Islands and the Bismarck Archipelago. In late winter and early spring large numbers fly south across Torres Strait and other waters to breed mostly on offshore islands of northern Australia, from the Kimberleys to about Mackay.

Last century and to a lesser degree in Banfield's day, its flocks were often immense. But in much of coastal Queensland, the clearing of coastal rainforest for sugar cane and farms has greatly reduced available food and shooting (now illegal) on the breeding islands has decimated the once vast companies of pigeons.

They come from the north in their thousands . . .

No birds of the air which frequent these parts attract more attention than the white nutmeg or Torres Straits pigeons . . . which resort to the islands during the incubating season.

White with part of each flight feather black, . . . it is a handsome bird, strong and firm of flesh, and possesses remarkable powers on the wing.

Half of the year is spent with us. They come from the north in their thousands during the first week of September, and depart during March. While in this quarter they seek rest and recreation, and increase and multiply on the islands, resorting to the mainland during the day for food.

Their flights to and from are made in companies varying from four to five to as many as a hundred—but the average is between thirty and forty. Purpose and instinct guide them to certain islands, and to these the companies set flight.

Towards the end of the breeding season, when the multitude has almost doubled its strength by lusty young recruits, for an hour and more before sunset until a few minutes after, there is a never-ending procession from the mainland to the favoured islands—a great, almost uncountable host.

Soon some of the tree-tops are swaying under the weight of the masses of white birds, the whirr and rush of flight, the clacking and slapping of wings, the domineering "coo-hoo-oo" of the male birds and the responsive notes of the hens; the tumult when in alarm all take wing simultaneously and wheel and circle and settle again with rustling and creaking branches, the sudden swoop with whistling wings of single birds close overhead, create a perpetual din.

Then as darkness follows hard upon the down-sinking of the sun, the birds hustle among the thick foliage of the jungle, with querulous, inquiring notes and much ado. Gradually the sounds subside, and the subdued monotonous rhythm of the sea alone is heard. . . .

E. J. Banfield

The dugong, *Dugong dugon*, or 'sea-cow' is one of those creatures many people have heard about—and that's about all. Briefly, a dugong is a large, air-breathing, vegetarian marine mammal, whose remote ancestors came from the land.

A large dugong may grow to three metres long. It has the shape of a plump torpedo with large, paddle-shaped pectoral fins and a horizontal tail like a whale's flukes to drive it forward with powerful up and down strokes. At the other end is a rather plain face whose main feature apart from smallish eyes is a very large, broad flexible upper-lip used for gathering seagrass* on which dugongs feed almost exclusively.

Dugongs are social animals, living in small parties to large herds. Their daily movements to suitable feeding grounds are dictated by the tides. They are peaceable, worthy creatures.

Unfortunately they are also easy to shoot and delicious to eat, and although widely distributed in the tropical and subtropical western Pacific and the Indian oceans, their world stronghold now seems to be the northern coasts of Australia, from Moreton Bay in the east round to Shark Bay in the west. Here they inhabit sheltered bays and inlets and the lee side of islands and other situations where seagrass grows luxuriantly.

In Banfield's time on Dunk Island, dugong were still abundant in the shallow coastal waters. They are still there, in yet smaller numbers. Only the provision of spacious marine parks round our northern coastlines and environmental controls to prevent such large-scale collapse of seagrass as we have recently seen in Victoria's Western-port will ensure their future.

* 'Seagrass' is composed of several genera of flowering land plants, notably *Zostera*, which have become adapted to life in shallow seas.

Dugong, *Dugong dugon*

. . . great . . . sportful water-babes

Dugong . . . still frequent these waters . . . Even in the narrow limits of Hinchinbrook Channel, through which the passing of steamers is of everday occurrence, they still exist, though not in such numbers as in the early days.

It would seem that the waters within the Great Barrier Reef may long continue one of the last resorts of this strange, uncouth, paradoxical mammal.

Half hippopotamus, half seal, yet in no way related to either, . . . the dugong is a herbivorous marine mammal, commonly known as "the sea cow", because of its resemblance in some particulars to that useful domesticated animal . . .

But, unlike the cow, the dugong has two pectoral mammae instead of an abdominal udder, and like the whale is unable to turn its head, the vertebrae of the neck being, if not fused into one mass, at least compressed into a small space . . .

When the mother is nursing her child, holding it to her breasts, she is careful as she rises to breathe, that it, too, may obtain a gulp of fresh air, and the two heads emerging together present a strangely human aspect . . .

Reddish grey, sometimes almost olive green in colour, with white blotches and sparse, coarse bristles, the animal has no comeliness, and yet when a herd frolics in the water, rising in unison with graceful undulatory movements for air, and the sunlight flashes in helioscopic rays from wet backs, the spectacle is rare and fine.

Rolling and lurching along, gambolling like good-humoured, contented children, the herd moves leisurely to and from favourite feeding-grounds, occasionally splashing mightily with powerful tails to make fountains of illuminated spray—great, unreflecting, sportful water-babes.

Admiration is enhanced as one learns of the affection of the dugong for its young and its love for the companionship of its fellows. When one of a pair is killed, the other haunts the locality for days. Its suspirations seem sighs, and its presence melancholy proof of the reality of its bereavement . . .

E. J. Banfield

The Beachcomber gives here a splendid picture of a huge strangler fig, *Ficus* sp., of the rainforest, full of ripening fruit.

Many rainforest birds feast on figs but most splendid of them all are the specialised fruit-pigeons, whose gapes can spread wide for swallowing largish fruits and whose strong feet grip like those of parrots for efficient climbing among the branches.

Unlike many pigeons, the colours of fruit-pigeons are rich and contrasting: deep glossy greens, pinks, purples, bright orange, soft yellow and soft grey.

Despite this, even the large, magnificent or wompoo pigeon, *Ptilinopus magnificus*, is very hard indeed to see when quietly feeding high in the canopy of a huge fig.

More brilliant still, the tiny purple-crowned and red-crowned fruit-pigeons, *Ptilinopus superbus* and *P. regina*, are almost impossible to find as they clamber in the leaves or make short wing-whistling flights between branches of sunlit leaves.

These, with the other birds mentioned and the handsome grey goshawk, *Accipiter novaehollandiae*, circling above, make up an essentially tropical Australian vignette. It is utterly different from the parched deserts that Sturt and Giles knew but strongly characteristic of the Australian region nonetheless.

Note: Because many of the bird-names Banfield used have been superseded, present-day names have been substituted in his text.

From [sunrise] to [sunset] birds feast and flirt . . .

The tyrannical fig-tree . . . in full fruit—pink in colouring until it attains purple ripeness—attracts birds from all parts, and for nearly a quarter of the year is as gay as a theatre.

From [sunrise] to [sunset] birds feast and flirt with but brief interludes.

A general dispersal of the assemblage occurs only in the tragic presence of a [goshawk] whose murderous deeds are transiently recorded by stray painted feathers.

But the fright soon passes, and the magnificent fruit-pigeon—green, golden-yellow, purplish-maroon, rich orange, bluish-grey, and greenish-yellow, are his predominant colours—resumes his love plaint in bubbling bass.

"Bub-loo, bub-loo, maroo," over and over again, in unbirdlike tone, without emphasis or lilt. "Bub-loo, bub-loo maroo," a grievance, a remonstrance and a threat in one doleful phrase; but to the flattered female it is all compliment and gallantry. . . . the white-headed, the pheasant-tail, the gorgeous "superb", the peaceful dove and the red-crowned fruit pigeon—most timorous of the order—are regular patrons, and each of the family has the distinctive demeanour and note . . .

Purple-crowned fruit-pigeon, *Ptilinopus superbus*

Baldwin Spencer

Professor Sir Baldwin Spencer (1860–1929) was a remarkably productive, remarkably humane and able man. He came to Melbourne in 1887 from Oxford to establish a department of biology at the University of Melbourne, and remained its head for three decades.

He later became an anthropologist when he accompanied the Horn Scientific Expedition, mounted by the universities of Adelaide, Melbourne and Sydney, to Central Australia in 1894. His anthropological investigations and photography of the Aborigines of this area were to become his main life's work.

At Alice Springs he met F. J. Gillen, special magistrate and sub-protector of Aborigines. They were to collaborate in several works that became classics: *The Native Tribes of Central Australia* (1899); *The Northern Tribes of Central Australia* (1904); *Across Australia*, 2 vols (1912) and, long after Gillen's death, *The Arunta: a Study of a Stone Age People* (1927).

Spencer was also a skilled field naturalist who skinned and prepared many of the birds and mammals he collected. Extensive travels in inland and northern Australia at the turn of the century made him familiar with wildlife that, still common then, has all but vanished now before the affects of European settlement.

One such threatened native is the beautiful, almost unbelievable bilby or rabbit-eared bandicoot, *Macrotis lagotis*. Once widespread over two thirds of Australia, it is now pegged back to pockets of inhospitable desert country in far south-western Queensland and the western Northern Territory and adjacent Western Australia.

Spencer's account from *Across Australia* indicates it was a common, abundant animal in the Centre well into this century.

Greater bilby, *Macrotis lagotis*

. . . a very pretty, graceful and delicately coloured animal

Almost every native has one or more tassels, which are worn hanging down over the forehead or suspended from the waist girdle, and are made from the tail tips of the rabbit-bandicoot . . . [whose] popular name of "rabbit" is due to its large ears; otherwise, apart from the fact that in size it approximately resembles a rabbit, it has no resemblance to this animal at all. It has a very long, pointed snout, with numerous small front teeth and strong canines, and feeds on vegetables, insects, and grubs. The fur is long, silky, soft, and generally grey-coloured with here and there a rufous tinge, save on the under side of the body, where it is white. The ears are almost naked and consequently the blood vessels give them a pink tinge.

The most striking feature, however, and the one which gives the animal its value in the eyes of the natives, is the tail. The basal third is grey, the middle third black, and the terminal third is marked by a prominent crest of white hairs on the upper side. The natives only use the latter. They cut off the flap of skin which carries the crest and twist this round and round in such a way that it forms a little brush of long white hairs. Sometimes as many as twenty of these brushes will be tied together to form the tassel which, as also the little brush itself, the natives call "alpita".

In many parts of the Centre, the burrows which this bandicoot makes are very extensive, the animal living in colonies. Each burrow has an entrance two feet or more in diameter, and around this the sandy soil is raised into heaps. Judging by the supply of alpita in every native camp, this bandicoot, or "Urgatta" as the Arunta natives call it, must abound; and it must also be a prolific breeder, otherwise the constant depredations of the natives would have exterminated it.

Baldwin Spencer

From his base at Charlotte Waters on the overland telegraph line in the southern Northern Territory, Baldwin Spencer set about collecting birds, mammals and reptiles of that baking arid region.

Though he had not come to Australia until his late twenties, by the 1920s Spencer had become one of the most experienced Central Australian travellers.

One brought up in the benign, tranquil atmosphere of an English summer could scarce find a region so different: a region technically semi-desert, yet one that could overnight produce roaring floods of sufficient force to tear centuries-old river redgums from the ground, then dwindle away into the sand, leaving only pools.

Even less might he imagine a heat so fierce that within moments it exposed to it. In this incident described in *Across Australia*, is an abrupt reminder of the kinds of adaptation that animals must make to survive such climates.

The reptiles involved in this incident were a sleek, swift dragon-lizard, now known as *Lophognathus longirostris,* and a western blue-tongue skink, *Tiliqua occipitalis.*

Western bluetongue, *Tiliqua occipitalis*

. . . baked to death by the great heat of the sand

As soon as the natives had been "sent out bush" with instructions as to what I specially wanted to secure, Byrne and myself went out some twenty miles to the west to see the country over which the flood waters of the Finke had just swept . . .

It was probably in some flood such as this that Leichhardt's party disappeared from sight for ever.

It was very hot, so much so indeed that the heat of the sand penetrated through the soles of our boots. There was an extraordinary feeling of ruin and desolation. Everything was perfectly silent. Not a bird about, except a few kites—not even a crow.

Two natives that we had taken out with us were busy collecting anything that they could find, especially lizards . . .

One of the most interesting forms [was] a . . . very graceful, active [dragon-lizard] . . .

The difference between a warm-blooded and a cold-blooded animal was impressed upon me when one of the natives brought in a very fine example of a large Skink, *Tiliqua occipitalis*. It is covered with scales and has short, stumpy legs, so that the under surface drags along the ground when it moves, or at least keeps coming into contact with it.

Being busy with some other specimen I told the boy to wait and not to let it go. To my surprise he put it down on the sand, simply saying in his broken English, "Him no go far, quick fellow him been tumble down altogether." He made no attempt to hold it and I thought it would escape, but the boy knew more than I did, and after moving along slowly for three or four yards, it actually did "tumble down altogether"; it was dead, baked to death by the great heat of the sand because it had no way of keeping its body cool when the sand on which it lay was hot.

There was in this respect a great difference between the stumpy Skink and the graceful long-legged [dragon]. The former had to keep under shelter during the heat of the day; and the latter, running along on its hind legs, which it can do at a remarkable rate, so that its body does not touch the ground, is always in the open, except along the watercourses, where, if pursued, it runs for shelter into the flood wrack . . .

G. A. Keartland

English born G. A. Keartland (1848–1926) became a compositor with the *Age*, Melbourne, by trade but was always an ornithologist by preference. He formed an extensive egg-collection, served as president of the Victorian Field Naturalists Club and wrote many papers for its journal.

In May 1894, Keartland was appointed naturalist/collector to the Horn Expedition to Central Australia, a joint scientific venture by the universities of Adelaide, Melbourne and Sydney to survey the geology, botany and zoology of the Macdonnell Ranges and the Aboriginal peoples of the region.

Two years later he also accompanied the Calvert Expedition to inland Western Australia. Keartland's comments here about inland birds mark the beginning of a realisation that birds in arid Australia, especially the seed-eaters, have special strategies for finding and retaining adequate moisture.

The zebra finch, *Poephila guttata*, is now known to be superbly adapted to life in the arid region. While most songbirds drink a sip at a time, the zebra finch puts its bill under and sucks like a pigeon, taking on a full load quickly before the flock is scattered by a butcherbird or sparrowhawk, which haunt the waterholes.

Once it has moisture, the zebra finch hordes it. The walls of its excretory tract can absorb the last trace of moisture from its droppings before they are voided. And in the last resort it can exist on shrinking water-sources that have become quite brackish to human taste.

European settlement, by opening woodlands and providing countless new sources of water, has been a boon to the zebra finch, whose total numbers across inland Australia must now be vast.

But we still do not know if they can 'smell water', as Keartland claimed in this paper from the *Victorian Naturalist*, vol. 7 (1901).

Zebra finches, *Poephila guttata*

Wherever water was found in the desert . . . these birds were seen in immense flocks . . .

I suppose it would be impossible to name a bird which consumes so much water in proportion to its size as this finch.

Wherever water was found in the desert of either Central or North-West Australia these birds were seen in immense flocks, and the more isolated the water the more numerous the birds.

I only saw one flock over 8 miles from water. They were very thirsty and tired, and settled on the tree under which we were having lunch, about 15 miles west of Joanna Spring. The billy was boiling on the fire, and the finches flew to the water casks, and, after hopping about them for some minutes, returned to the tree.

Immediately a pannikin of water was placed on the ground they drank ravenously, and then started off in the direction of a well which we afterwards found.

What attracted those birds to that particular tree when there were many others near? It must have been either the sight of the water in the billy, or the scent of the contents of the casks. . . .

James Dawson

James Dawson was a 19th-century grazier who farmed near Camperdown in the Victorian Western District. He was also a friend and champion of the Aborigines, and recorded all he could of their old ways while opportunity lasted.

Dawson's reference to a burrowing animal called a yaakar raises an important point to remember in any evocation of pre-European Australia: that the present distribution and diversity of native marsupials is very different from that known in the early days of settlement.

Victoria then had rock-wallabies and pademelons. The beautiful toolache wallaby (now wholly extinct) lived in the border region with South Australia. On the northern plains was a wombat and a hare-wallaby, and there were three attractive rat-kangaroos or bettongs—one of which, the burrowing bettong (*Bettongia lesueur*), although brown in colour and not black, may have been the 'yaakar' of Dawson's Aborigines.

These and a dozen species of other mammals in eastern Australia collapsed before the onslaught of clearing and grazing, the introduced cat, the fox, the rabbit and apparently from at least one serious epidemic disease. Even the red-necked wallaby, *Macropus rufogriseus*, though still present across central Victoria, is much reduced.

Unlike the Aborigines, we did not exercise 'a wise economy' in killing these animals. Intentionally or otherwise, we obliterated them, expunged them, from our part of the continent.

This excerpt is from Dawson's book *Australian Aborigines* (Melbourne, 1881).

... they eat the several kinds of kangaroo

The aborigines exercise a wise economy in killing animals. It is considered illegal and a waste of food to take the life of any edible creature for pleasure alone, a snake or an eagle excepted. Articles of food are abundant, and of great variety; for everything not actually poisonous or connected with superstitious beliefs is considered wholesome . . .

Of quadrupeds, they eat the several kinds of kangaroo, the wombat—which is excellent eating—the bear, wild dog, porcupine ant-eater, opossum, flying squirrel, bandicoot, dasyure, platypus, water rat, and many smaller animals.

Before the occupation of the great plains by cattle and sheep, there were numerous black and brown quadrupeds, called the yaakar, about the size of the rabbit, and with open pouches like the dasyures.

They were herbivorous, and burrowed in mounds, living in communities in the open plains, where they had their nests. They had four or five young ones at a time; and, from what the natives say about the numbers that they dug up, they must have furnished a plentiful supply of food at all times . . .

OPPOSITE: Red-necked wallaby, *Macropus rufogriseus*

138

Archie Campbell

About a century ago, the vast lowland rainforests of coastal northern New South Wales were known as the Big Scrub. As the Scrub was progressively felled for dairy farms there was immense loss of plants and animals that lived in it and in the priceless timbers that went up in smoke.

Today there are only meagre pockets of the Scrub left, here and there, a notable example being Stott's Island in the Tweed River beside the Pacific Highway, downstream of Tumbulgum.

Even these pockets are so fascinating and so rich in life that you can spend hours of delight exploring them, especially if you come from the south or from Europe, where such teeming diversity of colour, form and function is unknown. You are assailed by new scents and colours. In an astonishing parade, new birds and insects flash into view.

How much more impressive must Melbourne naturalist Archie Campbell have found it all in 1900, on his first visit north, when parts of the Scrub were still intact and he could stay in a quiet farmhouse close to it all.

His sense of freshness and discovery of a new world rivalled that of any Sturt or Oxley. The butterfly he describes is the gorgeous Richmond birdwing,* *Ornithoptera richmondia*, one of the largest Australian butterflies. It is still found on margins of rainforest, but these are much-reduced.

The South American *Lantana camara* is now a well-known pest in the region. This excerpt is from Campbell's article in the *Victorian Naturalist*, vol. 17 (1900).

* The Cairns birdwing, illustrated, though larger than the Richmond River form, is similar in general appearance.

OPPOSITE:
Cairns birdwing, *Ornithoptera priamus*

Cairns birdwing, *Ornithoptera priamus*

Butterflies are seen in myriads . . .

The "Big Scrub" grows in a rich red soil, the main tract extending from Lismore on the south of the Macpherson Range on the north, and the whole of the country is a delightful series of hills and hollows, with creeks and watercourses in abundance. But the scrub is now falling fast before the selector's axe, and dairy cattle in great numbers are thriving upon the rich pastures which take its place.

Butterflies are seen in myriads on a bright day, with gay colouring and quick flight searching in and out among the blossoms. The . . . Richmond River district can claim a species peculiar to itself, *Ornithoptera richmondia*, a large insect, measuring from 4½ inches across . . . named from its heavy flight, which is supposed to resemble that of a bird.

The female is quite a common object during the early summer months, . . . feeding with hundreds of other smaller butterflies, on some flowering scrub tree; but the male does not usually put in an appearance until late in December, when an exceptionally hot day will free them all from their chrysalids, hanging suspended among the creepers or in the branches of the trees, and on the morrow their dazzling green and black forms are seen everywhere.

This is the conclusion I came to from my own experience, for on New Year's Day there was a lull in the rains and the day dawned fair; the sun soon enveloped the place in a steaming heat, the thermometer registering 98° in the shade.

Next morning two strange butterflies were reported to me in the garden; they were . . . the first I had seen of the beautiful male, for previously none had been about. But shortly after, when I paid a visit to the Lantana bushes by the roadside, I was met with a sight I shall never forget.

I shall simply say 18 males were captured within the first half hour . . . They were beautiful and perfect specimens in the morning, but before the day was over all showed frayed wings to a more or less extent, for the contact with the plants and flowers soon destroys their delicate beauty . . .

Mrs Aeneas Gunn

Mrs Aeneas Gunn, *née* Jeannie Taylor, was a small energetic woman of free spirit. Born in Melbourne in 1870 she matriculated at the University of Melbourne and in 1889, with her sisters, opened a private school at Hawthorn.

She then became a visiting teacher, and at that stage of her life met Aeneas Gunn, a librarian and journalist, who had spent some years exploring north-western Australia and establishing pastoral properties.

In 1901 Gunn became a partner and manager of Elsey cattle station on the Roper River, 500 kilometres south-east of Darwin. Early in 1902, after their marriage in Melbourne, Aeneas Gunn and Jeannie travelled to Port Darwin by ship, then to remote Elsey.

After the death of her husband just thirteen months later, Jeannie returned to Melbourne, but her letters and stories of life at Elsey so captivated friends that they persuaded her to write the story.

Published in 1905, *The Little Black Princess* and its unforgettable young rogue of a character, Bett-Bett, became a minor classic. *We of the Never-Never* (1908), a re-creation of life at Elsey, made her famous. The books showed Australians a world unknown in southern cities. This excerpt is from *We of the Never-Never*.

Jeannie's animals were probably little red flying-foxes, *Pteropus scapulatus*, heading out of their camp at dusk to drink and feed. This small flying-fox is our most wide-ranging species, following the blossoming of eucalypts and melaleucas and the ripening of native fruits, especially figs, over much of northern and eastern Australia.

Perhaps less abundant than formerly, it still forms huge summer camps in northern riverside scrub, mangroves and rainforest pockets. Some camps attract hundreds of thousands.

Little red flying-foxes, *Pteropus scapulatus*

. . . a shrilling, moving cloud

Then, supper over, the problem of watering the horses had to be solved. The margins of the lagoons were too boggy for safety, and as the horses, fearing [crocodiles] apparently, refused the river, we had a great business persuading them to drink out of the camp mixing dish.

The sun was down before we began; and long before we were through with the tussle, peculiar shrilling cries caught our attention, and, turning to face down stream, we saw a dense cloud approaching—skimming along and above the river: a shrilling, moving cloud, keeping all the while to the river, but reaching right across it, and away beyond the tree tops.

Swiftly it came to us and sped on, never ceasing its peculiar cry; and as it swept on, and we found it was made up of innumerable flying creatures, we remembered Dan's "flying foxes"

In unbroken continuity the cloud swept out of the pine forest, along the river, and past us, resembling an elongated kaleidoscope, all dark colours in appearance; for, as they swept by, the shimmering creatures constantly changed places—gliding downwards as they flew, before dipping for a drink, to rise again with swift, glancing movement, shrilling that peculiar cry all the while.

Like clouds of drifting fog they swept by, and in such myriads that, even after the Maluka began to time them, full fifteen minutes passed before they began to straggle out, and twenty before the last few stragglers were gone.

Then, as we turned up-stream to look after them, we found that there the dense cloud was rising and fanning out over the tree tops. The evening drink accomplished, it was time to think of food . . .

A. G. Bolam

People stationed in outback parts of Australia often had, and in parts still have, the chance to see the land much as the explorers did and know its wildness and integrity.

In some ways they were better placed: undriven by the need to keep moving, stationed in one place for years, they knew the best local habitats and had the matchless advantage of being able to watch the routines of wildlife through seasons good and bad.

A. G. Bolam, stationed at Ooldea on the transcontinental railway in south-western South Australia in the early years of this century, had such opportunity.

Those who have ever searched for hopping-mouse burrows will realise why the quest is often maddeningly unproductive. They will also admire Bolam's persistance and powers of observation.

Mitchell's hopping-mouse, *Notomys mitchelli*, is distributed over parts of arid southern Australia and is now uncommon. It is one of a small group of native rodents which have developed the shape and the efficient hopping techniques of the kangaroos, to which they are entirely unrelated.

This is convergent evolution: the coincidental development of similar strategies and physical form and function by unrelated animals and plants. Hopping-mice also dig well and can actually manufacture much of the water they need from carbohydrates in the seeds they eat. They are beautiful, impressive little animals.

This excerpt comes from Bolam's much-admired small book, *The Trans-Australian Wonderland*. First published in Melbourne in 1923, it went through five editions— remarkable success for a book of its nature. A sixth, facsimile, edition was issued in 1978 by the University of Western Australia Press.

. . . he hops . . . at an incredible speed

The Kangaroo Mouse . . . known to the blacks as Ool-git, is found well distributed all over the countryside. He . . . derives his . . . name from the . . . explorer Sir Thomas Mitchell, who was the first to describe him . . . He stands on the hind legs, and, although no bigger than the ordinary domestic mouse, he hops along at an incredible speed—so fast, in fact, that a good dog cannot overtake him. He hops three feet at each bound, and can continue to do so as long as pursuit lasts . . .

The Kangaroo Mice live in burrows, which are very cunningly constructed, a peculiar feature being what is commonly known as the dummy hole. A burrow is started, and all the earth is very carefully scratched out at the one and only opening. To do this work two mice are generally employed, one behind the other; and when the front mouse starts burrowing, the other one, which is immediately behind, scratches the dirt away . . . As the excavation progresses, one or two holes make their appearance some distance away from the original opening, and, strange to say, no earth is scratched out at any of the new openings, but it is all taken out at the first entrance.

When the burrow is finished, the original hole is completely blocked up by the dirt which has been stacked outside, and it is used no more. In the course of time, this dirt is blown away or covered over with leaves, and assumes a very aged appearance. This is just what the mice want, as the ground, being so obviously unused, is passed over without notice.

The holes, therefore, become very difficult to detect; and a small opening of three-quarters of an inch, in all probability under a bush, is the only entrance to their underground home . . .

Mitchell's hopping mouse, *Notomys mitchelli*

Erwin Nubling

In general there have been two stages in the matter of Europeans getting to know Australian wildlife. The first stage was that of discovery and initial reaction, often one of surprise, even perhaps of delight, that something so utterly new, like a tree-kangaroo or a platypus, could exist.

The second stage came a century or more later, when investigators, professional zoologists or interested amateurs began to look objectively at particular animals.

So it was with bowerbirds. The first European observers, once they knew what the bowers were, saw them in human terms as 'playgrounds'. Even the perceptive John Gould, who prided himself on making our bowerbirds known to the world, shared this view.

It was only much later, when something was known about the phenomenon of bird territoriality and display, that a new view arose about bowerbirds.

A pioneer in such investigations was Erwin Nubling, son of an Austrian civil engineer and railroad builder. Nubling came to Australia in 1907 and made a career in the steel industry, eventually becoming managing director of a Sydney steelworks.

The careful study of satin bowerbirds, *Ptilonorhynchus violaceus*, in Sydney's Royal National Park was to him a leisure-time delight. Nubling's native gift for accuracy helped establish an important principle about bowerbird display: that the decorative objects in the bower are *not* simply playthings, nor simply objects expressive of an aesthetic sense, though they are perhaps that as well.

Before Nubling, it was generally accepted that bowerbirds are attracted by any bright object. Nubling's contribution, published in the ornithological journal the *Emu*, vol. 21 (1921), gained from observations at many bowers, was that male satin bowerbirds chose predominantly

Satin bowerbird, *Ptilonorhynchus violaceus*

blue play objects or those of a particular pale yellowy green.

These are the colours of the plumage, eyes and bill of the male satin bowerbird itself and it took a brilliant zoologist, Professor Jock Marshall, who was growing up in Sydney when Nubling was making his observations, to demonstrate the significance of *that* in *Bowerbirds. Their display and breeding cycles*. (Oxford, 1954).

Marshall's work showed that the feathers, flowers, fruits and shells (or near human habitation, blue ribbons, 'blue-bags', blue glass, plastic, metal or paper) a bowerbird brings to his bower are objects of extreme fascination to this aggressively territorial bird. One could suggest they become tokens representing the rival males.

Territorial song and display in birds has several functions: to warn off rival males, announce claim to a certain territory, attract a female or females and—as weeks pass—prepare their reproductive systems for mating, nest-building, egg-laying and young-rearing. (Unlike humans, most birds are fertile for one period only each year. Outside that period their reproductive systems 'regress'. It takes many factors, internal and external, to re-stimulate them and synchronise the preparedness of males and females.)

The enhanced display of the male bowerbird seems to achieve this, often without need for physical combat between rival males. Combat seems to have been part-replaced by formalised and stylised display, for the greater good of the species as a whole.

It must have given Erwin Nubling, migrant from turbulent Europe, some pleasure to have had a hand in that discovery.

. . . a decided preference for blue

In October, 1920, in the National Park, south of Sydney, I . . . was pleased to meet the Satin Bower-Bird in its natural haunts, which had long been my wish . . .

On the 10th of October, we saw a blue-black Satin [bird] fly over the road. It alighted first in a tree, and then dropped to the ground in a small patch of bracken ferns, near the water's edge. When the bird flew, we found the bower . . .

It was built at the edge of the bracken, some of the fronds bending over it. The platform on which it stood was . . . constructed neatly of long, dry she-oak (*Casuarina*) needles, with a few thin sticks between them, looking almost like a carpet.

On the platform, more on the sides than in front, we found many dry leaves of *Banksia serrata*, . . . Besides, there were snail shells of two kinds, a piece of string, pieces of blue paper, two small bags of washing blue, and some blue glass . . .

The walls [of the bower] were but little arched on the top, and the passage wider at the top than at the bottom, where the sticks curved in towards the centre . . .

After waiting a while . . . we saw the bird in a tree overhead, with a stick in its bill. It flew first down to the stump, and then to the platform, where it hopped about, emitting repeatedly a buzzing sound, and uttering short cries, whilst its wings were slightly raised . . .

It is interesting to see him working at the walls, lifting a stick out at the end of a wall, turn his head round, raise it and give it a twist, and then with his bill ram the stick down in another place . . . All . . . done in a thoughtful manner without haste . . .

On Saturday, 30th October, bower No. 1 . . . was as before, but for an extra supply of many pieces of blue glass and fresh *Billardiera* blossoms . . . At bower No. 2 our winged friend was soon heard and detected. A piece of blue velvet was the only novelty in his collection . . .

At 4.15 next morning I was up listening to the manifold voices of the birds in the gully from early dawn to sunrise, starting one after the other with their song. About an hour later we could, from above, see the Satin [bird] about his bower. Slipping to the bower during our early breakfast, I noticed the [bird] nip off a stalk of blue *Dianella* blossoms and deposit it on the platform . . . After 7 he brought two puff balls, some flowers having previously been deposited. His last morning visit was after 9, when the bower was becoming sunny . . .

[Then] he sat on a log, four feet away . . . when I could once more at close quarters admire his beautiful [blue] eyes and glossy plumage, as smooth and sleek as the skin of a seal . . . His beak is of a rather light yellowish or whitish-green, not unlike that of the blossoms he is so fond of . . .

The birds of both bowers show a decided preference for blue and yellowish-green, as regards their decorations, the only exceptions being perhaps the brownish snail shells, yellowish-brown Cicada larval shells, and the more olive green puff-balls. A thing of another colour is, as a rule, rejected.

I do not intend to generalise from this, but whereas in nearly all descriptions I have read of play-grounds, it is implied that anything bright is regarded by the Satin Bower-Bird as a fit object for collection, my observations so far do not bear out this contention . . .

Straw-necked ibis, *Threskiornis spinicollis*

W. H. D. Le Soeuf

W. H. D. Le Soeuf (1857–1923) was a member of an Australian family prominent in natural history, particularly in connection with zoological gardens. He was also a grandson of the early, recently-celebrated bird-painter John Cotton, who settled near Mansfield, Victoria, in the 1840s.

Le Soeuf was particularly interested in birds. He gave long service to the Royal Australasian Ornithologists' Union and travelled widely. Visiting the New South Wales Riverina near Deniliquin in October 1900, Le Soeuf came on an enormous nesting colony of straw-necked ibis, *Threskiornis spinicollis.*

This ibis feeds in dry grasslands as well as swamps and the formation of a colony of this size probably depended on the presence in the district of big numbers of plague locusts, *Chortoicetes terminifera.*

Le Soeuf's figures not discounted, it is now thought that while enormously beneficial, ibis probably do not fully live up to their almost legendary reputation for controlling locust plagues. They are probably most valuable when locust numbers are down, when the quantities they destroy may be significant.

That apart, the gathering of such enormous numbers of ibis to breed and their wavering formations across the open sky of the plains was a spectacular demonstration of the wealth of the Australian natural world, not easily forgotten. But such numbers are seldom seen today.

Le Soeuf's excerpt is from the *Victorian Naturalist*, vol. 7 (1901).

Straw-necked ibis, *Threskiornis spinicollis*

. . . the immense utility [of] these birds

The most interesting sight seen was a colony of the Straw-necked Ibis, . . . nesting in a swamp of about 600 acres, which was covered with Lignum *(Muehlenbeckia cunninghamii)* bushes from six to ten feet high, and in which the water was from two to three feet deep . . .

As the swamp was approached, a curious sound, something like breaking surf on the shore, was heard, caused by the immense numbers of birds flying about and emitting their hoarse cry; but comparatively few birds were seen flying above the Lignum, and one could not tell that such a vast host of birds were nesting there. . . .

After having been all through the swamp and carefully noted the numbers on a small area, . . . [we] came to the conclusion that the minimum number was 200,000 . . .

No illustration . . . would give much of an idea of the number of birds or extent of the rookery, unless a photo. was taken from a balloon or some such eminence, for even when a gun was fired off only those in the immediate vicinity—say, of sixty yards—would rise, and not even all those unless they caught sight of the intruder; they would then circle round at a considerable height up, but would soon settle again . . .

The . . . food consists of grasshoppers, caterpillars, freshwater snails; &c., and if the young birds are handled much they occasionally eject the food from their stomach.

The contents of an average crop of an adult bird contained by actual counting 2,410 young grasshoppers, five freshwater snails, several caterpillars, and some coarse gravel which if you multiply by 200,000, brings up a big total of 480,000,000 odd grasshoppers, . . . as well as vast numbers of caterpillars and snails, and also these latter are the host of liver fluke, which sheep so easily get in certain damp localities . . .

One must remember that this is going on every day, so a little idea can be formed of the immense utility these birds are in destroying noxious insects. Then, again, the average number of young is about 2½ to each pair of parent birds, and the contents of their stomach must reach an enormous total, as they all seemed gorged with food. . . .

Thomas Austin

Deprived of habitat by clearing, settlement and the effects of grazing by stock and by rabbits, decimated by foxes and guns, the lordly Australian bustard, *Ardeotis australis*, once abundant and widespread over the whole continent, is now nearly extinct in settled south-eastern Australia. In Victoria, the chances now of seeing a bustard anywhere but on a few properties in the thinly populated Mallee region are remote.

Yet less than a century ago, as this account recalls, bustards still lived and bred on the plains of Victoria's Western District nearly to the outskirts of Geelong and even Melbourne. In our civilisation, it is unlikely that wild bustards will ever be seen there again.

Thomas Austin (1874–1937) was a son of Thomas Austin of Barwon Park Station near Winchelsea, Victoria, remembered for his successful introduction of the rabbit to Australia in 1859.

From his property near Cobborah in New South Wales, the younger Thomas collected birds' eggs and corresponded with leading birdmen like Gregory Mathews and also A. J. North, ornithologist at the Australian Museum. Austin's fascinating historical note on bustards near Geelong was quoted by North in his *Nests and Eggs of Australian Birds* (Sydney, 1901–04).

. . . *Bustards in Victoria grow to a great size*

Bustards were [formerly] fairly plentiful in the Western District of Victoria, especially on Avalon Station, twelve miles north from Geelong, this being the only place I have ever found their eggs.

During October, 1889, while riding at a canter with one of my brothers through a large open paddock, we flushed a Bustard from her nest, in which was a single egg.

We were riding about six feet apart, and as the bird flew from the nest when directly between two horses, her out-stretched wings were beneath their noses; upon looking down I saw the egg, which was quite fresh . . .

In the same paddock I knew of another nest with a single egg. This was on the side of a stony rise, and if approached from the north I could get fairly close to the sitting bird before she could see me, but coming towards the nest from a southerly direction, she could easily see me a mile away, and would leave the nest and sneak away when she was only just visible to the naked eye, even when I knew where to look for her . . .

Male Bustards in Victoria grow to a great size. I have shot them as much as thirty-two pounds in weight (in the feathers) but in my life time were never in very great numbers, but on Murdeduke Station, near Winchelsea (where they were strictly protected), I once counted eighty-three within a mile of the homestead, many of them being just outside the garden fence . . .

Australian bustard, *Ardeotis australis* in territorial display

Australian bustard, *Ardeotis australis*

David Fleay

David Fleay has long been regarded as one of the mainstream pioneers in the making known of Australian wildlife. Although born in 1907, after the work of the more conventional explorers was done, Fleay has used his long lifetime to explore the life histories of Australian animals. He began keeping and studying them as a schoolboy in Ballarat, Victoria; then successively as curator of the Australian section at Melbourne's Zoological Gardens, director of the then Sir Colin Mackenzie Sanctuary at Healesville, Victoria, and finally at his own Fauna Reserve at West Burleigh in Queensland.

The 'plain tale' of the living, feeding, breeding habits and psychological needs of many Australian animals: platypuses, wombats, numbats, native cats, wedge-tailed eagles, barking owls and many others have been laid bare by Fleay's efforts.

Fleay participated in the first successful 'milking' of a taipan, at a time when there was no known antidote for its bite.

He was the first to coax the platypus and many another elusive or touchy species to breed in captivity, including his great love, the lordly powerful owl, *Ninox strenua*.

It was David Fleay who first began to make known the ways of this mysterious, imposing, golden-eyed dweller in Australia's south-eastern forests and its spitfire cousin, the barking owl, *N. connivens*.

David Fleay's compelling narrative style makes clear his own love for these great birds, their wild neighbours and the country they inhabit.

These excerpts from his book *Nightwatchmen of Bush and Plain* (Jacaranda, 1968) tell of his early wanderings in the Korweingeboora Forest, east of Ballarat, when 'the world was full of wonder and delight'.

. . . a colossal, aristocratic, altogether magnificent . . . bird

Summer time, when we were lucky enough occasionally, to borrow an old grocer's van and still more ancient, jibbing horse with a foul breath called 'Bill', we roasted in the blasting heat of north wind days or choked in bushfire smoke; but through it all, no one else ever intruded upon our private world of fact and fantasy, and a romance and strangeness beckoned always.

The second trip was fruitless as far as discovering the origin of the mysterious call was concerned, though it came to us during the night, haunting and remote far up on the ridge tops—particularly in the stillness of the earliest signs of dawn.

However, on the third expedition, immediately prior to picaninny daylight, that elusive voice began in the same offshoot gully whence it had originally arisen. It was no trouble to get going as we slept—full marching order—in 'boots and spurs', hat and jacket.

Following a quiet stalk, I eventually got close enough to realize that the deep 'woo-hoo' was emanating from upper branches of a large blackwood tree right in the floor of the gully. Then came the realization in growing daylight that I was gazing directly at a now perfectly quiet, and entirely motionless, giant of an owl.

I'd never seen anything larger than the little Boobook, but this enormous, upright, brown fellow with strongly barred chest and abdomen, staring haughtily and unblinkingly at me from great orange eyes, seemed at least two feet high. He was the most aloof impressive thing of his kind I had ever seen.

As I moved slowly below the tree, a piercing stare followed my every movement. Below him on the ground were innumerable bones and quite a collection of fresh, parcelled-up fur pellets, obviously disgorged from meals over a lengthy period.

This then, was a regular perch. No other bird was seen and on other visits of inspection, this acquaintance of future seasons was not always there. However, for the first time in my life, I'd come across a Powerful Owl in its natural state and the event was momentous. Never had I seen such a colossal, aristocratic and altogether magnificent hunting bird.

To find out more about Powerful Owls was bird-observing under difficulties. Remember, we had first of all to push cycles for thirty-five miles, lugging them in the later stages along rough timber-strewn bullock tracks to reach our operational area.

But that was the beauty of trips in those days. We were part of our surroundings and everything about us had its particular impact. The world was full of wonder and delight . . .

Powerful owl, *Ninox strenua*

PICTORIAL ACKNOWLEDGEMENTS

Graham Pizzey and the publishers would like to express their gratitude to the following photographers and organisations for permission to reproduce the photographs in this book, with special thanks to Mary Rose Gordon, archivist with the National Photographic Index of Australian Wildlife, the Australian Museum, Sydney, and to Daphne Keller, Australasian Nature Transparencies, Melbourne. Numbers listed below indicate the page numbers on which photographs appear.

AUSTRALASIAN NATURE TRANSPARENCIES   R. J. Allingham 12–13, C. A. Henley 15, A. Burbidge & J. Raines 16–17, M. F. Soper 23, Ron & Valerie Taylor 32, G. B. Baker 37, B. G. Thomson 52, Tom & Pam Gardner 55, 81, 146, 147, Kathie Atkinson 87, C. B. & D. W. Frith 90, 96, 97, 110, 144, R. J. Tomkins 94, Ralph & Daphne Keller i, 4, 128

HANS & JUDY BESTE   iii, 2, 8, 11, 20, 24, 25, 43, 51, 56–7, 67, 75, 76, 82, 83, 86, 91, 93, 104, 117, 118, 124, 125, 126–7, 133, 141, 142, 143

GRAEME CHAPMAN   9, 10, 19, 22, 30–31, 40–41, 68–9, 84 (above), 99, 102, 106, 136–7, 149, 151

NATIONAL PHOTOGRAPHIC INDEX   P. Watts 1, P. Klapstke 18, 42, C. A. Henley 26, P. G. Roach 74, D. & M. Trounson 84 (below), G. Suckling 89, G. Weber 103, J. Gray 105, B. Cropp 132, A. G. Wells 134, H. Elmann 135

GRAHAM PIZZEY   4, 5, 21, 35, 38, 44, 45, 47, 48–9, 58–9, 60, 63, 64, 65, 71, 73, 77, 78–9, 85, 100, 107, 112, 113, 119, 121, 129, 130, 140, 148

LEN ROBINSON   6, 7, 29

G. E. SCHMIDA   39, 109, 122

PETER SLATER   3, 133

GLEN THRELFO   114, 139

The coloured plate on page 70 is reproduced from John Gould's *Mammals of Australia* (London, 1845–1863)